The Human Enhancement Debate and Disability

# The Human Enhancement Debate and Disability

## New Bodies for a Better Life

Edited by

Miriam Eilers
*Ruhr-University Bochum, Germany*

Katrin Grüber
*Institut Mensch, Ethik und Wissenschaft, Berlin, Germany*

Christoph Rehmann-Sutter
*University of Lübeck, Germany*

palgrave
macmillan

First published 2014 by
PALGRAVE MACMILLAN

Palgrave Macmillan in the UK is an imprint of Macmillan Publishers Limited, registered in England, company number 785998, of Houndmills, Basingstoke, Hampshire RG21 6XS.

Palgrave Macmillan in the US is a division of St Martin's Press LLC, 175 Fifth Avenue, New York, NY 10010.

Palgrave Macmillan is the global academic imprint of the above companies and has companies and representatives throughout the world.

Palgrave® and Macmillan® are registered trademarks in the United States, the United Kingdom, Europe and other countries.

ISBN 978–1–137–40552–4

This book is printed on paper suitable for recycling and made from fully managed and sustained forest sources. Logging, pulping and manufacturing processes are expected to conform to the environmental regulations of the country of origin.

A catalogue record for this book is available from the British Library.

A catalog record for this book is available from the Library of Congress.

# Contents

## Part III Utopian Ideas and Real Embodiment

# Illustrations

# Acknowledgements

This publication was printed with financial support of the Federal Ministry of Education and Research of Germany.

# Foreword: Five Thoughts About Enhancement

## There is no single disability response to the enhancement debate

When faced with the difficulties of disability, the traditional answer has been medicalization. If only these difficult conditions could be cured or prevented, goes the argument, those poor benighted souls would be better off. I support medical research, and public health, and rehabilitation, and all the other clinical interventions which can improve health. But medicalization is not the full answer, and is often not the most appropriate and cost-effective answer, and sometimes is entirely the wrong answer.

The whole thrust of the disability rights movement over nearly 50 years has been to broaden our understanding of the disability question, and to show that the answers are also civil rights, barrier removal, inclusion in mainstream services, independent living, self-advocacy, and many other ways of accommodating difference. Medicine has its place, but alone it cannot solve the problem of disability, and sometimes can even make things worse.

This is why so many disability activists and disability scholars are so skeptical about the promise of enhancement technologies, be they genetic or pharmacological, surgical or prosthetic. Enhancement sounds very individualistic, it sounds very expensive – and hence unlikely to be available to all – and it sounds like it misplaces the disability problem.

However, this skepticism is not the only response to enhancement from the disability community. There are also those who say, from their position of difference, that we should embrace the prosthetic promise. If you are yourself abnormal in your embodiment, if you are used to surgical reshaping and orthotic or prosthetic or assistive technology solutions – wheelchairs and communication devices – not to mention having to pop pills every day to compensate and remedy and tweak, then it becomes much easier to envisage a different, wider, deeper, and more extensive engagement with enhancement. And when your whole society is relying on technological tweaks and boosts, then your own ventilator or anti-psychotic drugs or power chair perhaps becomes less noticeable, more acceptable.

## Start with an adequate understanding of disability

In my view, disability is the outcome of an interaction between the individual and a health condition, and the wider physical and social environment. This approach is compatible with both the World Health Organization International Classification of Functioning, Disability and Health, and the Convention on the Rights of Persons with Disabilities. Further, I believe that disability cannot be reduced either to a simple biomedical issue, or conversely to a purely social issue – whether the latter is conceived in terms of barriers and oppression, or in terms of social constructionism. Disability starts with bodies, and some of the disadvantage which disabled people experience derives from the decrements in functioning which flow from their health condition. But disability does not end with bodies because social arrangements, cultural representations, and public attitudes create additional burdens which make life with disability more difficult. So disability, on average, means having a harder life, because of the disadvantages which flow from the interaction of these less-than-usually-functional bodies and these discriminatory or neglectful societies. For me, there is little benefit in being relativist about disability. Disability may not be a tragedy, but neither is it just a neutral difference: disability, as I have argued elsewhere, is perhaps best conceived of as a predicament (Shakespeare 2013).

## The lived experience of disability challenges the quest for perfection

Despite the objective difficulties that confront disabled people – coping with a malfunctioning and possibly deteriorating body, and enduring the barriers and discriminations of an unwelcoming society – for most people, life with disability is quite tolerable. Evidence suggests that the subjective quality of life is rather high, certainly higher than non-disabled people imagine (Albrecht and Devlieger 1999; Amundson 2005). Health does not determine happiness, in most cases, although persistent pain, whether mental or physical, is hard to live with. Life with disability, in general, is entirely livable.

This last point is an important correlative to the more lurid exponents of enhancement, with the insatiable quest for stronger, fitter, faster, longer. Contra the perfectionist, it suggests that according to empirical evidence, less does not necessarily mean worse, and that perhaps more may not equate to better. The well-evidenced success of disabled people in leading good and full lives, notwithstanding their impairments and their experience of exclusion, provides evidence that a happy life does not require particular attributes. It may even be, although it is harder to cite proof, that being faced with obstacles and difficulties contributes to a more well-balanced and harmonious outlook. People can put things into proportion, they can see what is most important, perhaps they give up ceaseless striving and accept their situation. This has lessons for those who pursue the transhumanist chimera, but it seems unlikely that this will be acknowledged.

## Enhancement is about phantoms

I write this preface from Geneva, where Mary Shelley's *Frankenstein* came into being, the city of CERN and other cutting-edge science, the city of the historical League of Nations and the modern World Health Organization and many other visions of a better world, from Calvin onwards. This is a city of reality, but also of fantasy.

We are poised on the cusp of a future where we are promised transformations in human embodiment every bit as profound as the transformations in science, manufacturing, communications, travel, and understanding which emerged in the last century. None of us know whether this is sober prediction or science fiction. But I have a strong suspicion that the enhancement debate is about phantoms, about our secret hopes and fears and psychic troubles. We fear death, we daydream about physical and mental prowess, we want the very best for our children. These are normal human emotions, and it is hard to imagine that they will ever change. But these are not our only phantoms: we also dream of control and domination and superiority, fantasies which are more dangerous and more exclusive.

## The future is out of reach

When it comes to envisaging futures, fantasies are not a reliable guide. We need grounded, balanced, and immediate understandings. Disability research can offer us this correlative. In the *World Report on Disability*, to which I contributed, there is data about access to health and rehabilitation which shows how uneven the distribution of actually existing therapies and technologies is. Only 5–15 percent of disabled people worldwide have access to the assistive devices they need – and this refers to simple gadgets such as hearing aids, wheelchairs, crutches, and artificial limbs. According to the World Bank, two billion people in the world lack access to electricity. In the context of low and middle income countries, the solution is not the iBot, the futuristic power chair which lifts the owner to make eye contact, and enables them to negotiate stairs and other obstacles (cost $25,000). The solution is the Motivation basic wheelchair (cost $200). Rather than Oscar Pistorius' $13,000 state-of-the-art Icelandic blades, the option for an amputee in Sierra Leone or Haiti is more likely to be the Jaipur foot (cost less than $30). In this real world of seven billion human beings, people with epilepsy do not get the anti-convulsive drugs they require, people with depression lack access to proven treatments, and there is a lack of simple painkillers in many settings.

None of this means that enhancement technologies are intrinsically immoral or mistaken. But from the point of view of justice, it is clear that technologies have to be evaluated in terms of their potential for widening or narrowing inequality. Computers and space travel were pioneered at about the same time, and both challenged expectations and showed human ingenuity at its best. Computers are now cheap and ubiquitous, and it is hard to imagine life without them. Conversely, despite the best efforts of Richard Branson, even rich individuals are still not taking their vacation on the moon. Which is the better analogy for a particular enhancement technology?

Standards and expectations change, and sometimes technology does 'trickle down' from the elites to the masses. Mobile phones are a high-tech device which started as a pretentious luxury. But now, cheap and simple handsets have liberated farmers and fishers and traders in Africa, and brought the benefits of knowledge and communications to billions of people. Many enhancement technologies, I would argue, are unlikely to be similarly liberating.

It is an honor to be asked to introduce this wide-ranging collection of papers. Different disciplines are relevant to the topic of disability and the body. Disability studies has an important and valid contribution to make to the question of enhancement, and the voice of disabled people should be heard just as much here, as it is in other theoretical or policy debates. This book promotes the inclusive conversation and grounded analysis that this often exclusive and abstract topic lacks, and I welcome this sober and balanced contribution to an often over-heated field.

*Tom Shakespeare*
*World Health Organization*

## References

Albrecht, G. L. and Devlieger, P. J. (1999): The disability paradox: High quality of life against all odds. In: *Social Science and Medicine*, 48, pp. 977–988.
Amundson, R. (2010): Quality of life, disability, and hedonic psychology. In: *Journal for the Theory of Social Behaviour*, 40(4), pp. 374–392.
Shakespeare, T. (2013): *Disability Rights and Wrongs Revisited*. London: Routledge.

# Contributors

**Stuart Blume** has worked at the University of Sussex, the London School of Economics, the Organisation for Economic Co-operation and Development, and in various administrative positions, including in the Cabinet Office, London, and as Secretary of the Committee on Social Inequalities in Health (the Black Committee). From 1982 to 2007 he was Professor of Science and Technology Studies at the University of Amsterdam, where he is now Professor Emeritus. From 2009 to 2012 he was an advisor on bioethics to the World Federation of the Deaf. He is the author of *The Artificial Ear: Cochlear Implants and the Culture of Deafness* (2010) and co-author of *Protecting the World's Children: Immunisation Policies and Practices* (2013).

**Sigrid Bosteels** is a faculty member of the University College of West Flanders, Belgium, where she teaches philosophy and research methods. Trained in cultural sociology, she is currently working on a doctoral dissertation at the University of Ghent. Rooted in critical philosophical interrogations of the notion of normality and discussions regarding the predictability and malleability of lives, her research focuses on the legitimacy of early interventions in the lives of deaf children.

**Morten Hillgaard Bülow** has a background in history and philosophy/science studies and is currently a PhD candidate at the interdisciplinary Center for Healthy Aging, Faculty of Health Sciences, University of Copenhagen. He is based at the Medical Museion (a university museum for the history of medicine in Copenhagen) and his research interests include queer-feminist bioethics and materialities; the intersection of gender, disability, aging, and science; and the history of medicine in general. His current project investigates the history of the concept of successful aging within aging research, and discusses the scientific and ethical consequences of different understandings of aging. He has previously published about the historical treatment of individuals with XXY-chromosome variations in Denmark.

**Sascha Dickel** studied political science and sociology at the universities of Marburg and Frankfurt. From 2005 to 2010 he was a PhD student at the Institute for Science and Technology Studies at Bielefeld University where he also worked as a post-doc. In 2008 he was a visiting fellow at the Cardiff School of Social Science and in 2011 a Deutscher akademischer Austauschdienst (DAAD) research fellow at the American Institute for Contemporary German Studies at Johns Hopkins University, Washington DC. Since 2012 he has worked at the Institute of Ecological Economy Research in Berlin. His main research topics are technology assessment, the analysis of visions and scenarios and the study of socio-technical change. He also writes science fiction short stories and is blogging at remembertomorrow.net.

**Miriam Eilers** graduated in Medicine at the University of Lübeck in 2009 and is currently a member of the Mercator Research Group 'Spaces of Anthropological Knowledge' at Ruhr-University Bochum in Germany. Her research involves the interaction between bodies and society reflected in arts, literature, and science in the first half of the 20th century. Based on the work of the physician Fritz Kahn (1888–1968) as well as the Kahn archive, she is investigating the production and circulation of popular scientific texts as well as man–machine iconology before 1938. She is particularly interested in the iconography of popular scientific publications as well as transcription of knowledge in times of social upheaval. She has published about the history of enhancement in the Weimar Republic (2012).

**Lisa Forsberg** is a PhD student at the Centre of Medical Law and Ethics, King's College London. Lisa holds undergraduate degrees in political science and practical philosophy from Stockholm University and an MA in medical ethics and law from King's College London. She is also affiliated with the MIC Lab research group at the Department of Clinical Neuroscience, Karolinska Institute, Sweden.
Lisa's PhD thesis concerns public interest restrictions of the freedom of individuals to consent to controversial medical procedures, focusing in particular on procedures where neurotechnology is used.

**Trijsje Franssen** obtained her research master's degree in philosophy in 2006 at the University of Amsterdam. During her master's she completed an internship at the editorial office of the philosophical monthly *Filosofie Magazine*. After graduating, she worked at the University of Amsterdam, teaching modules in sociology of the arts, philosophy of culture and modern philosophy. In October 2009 she started her PhD at the University of Exeter under the supervision of Professors

John Dupré and Michael Hauskeller. Her thesis concentrates on today's debate on human enhancement. It focuses on the role of the ancient myth of Prometheus in the debate and, in particular, its relationship to (often implicit) moral claims and arguments on human nature. Apart from these subjects her research interests include disability, athletic enhancement, and posthumanism.

**Katrin Grüber** has been Director of the Institut Mensch, Ethik und Wissenschaft in Berlin since 2001. She was a member of the State Parliament of North Rhine Westphalia from 1990 to 2000. From 1995 to 2000 she was Adjunct Professor of Political Science at the Faculty of Philosophy, Heinrich-Heine-University, Duesseldorf and in 2001 Adjunct Professor of Political Science at the Institute of Nursing Science, University of Witten-Herdecke. Her research interests are the implementation of the CRPD (convention of the rights of disabled people), Science and Technology Studies (STS), and genetics. She is the co-editor of several books on ethics and disability.

**Kathrin Klohs** is currently writing a PhD thesis on fictional representations of academia at the science studies program at the University of Basel, Switzerland. Prior to that she studied German language and literature, philosophy, and classics in Freiburg, Germany. Her research interests also include sociology of literature and film studies.

**Nikolai Münch** studied political science, history, and economics at the Friedrich-Alexander-University Erlangen and at the University of Salamanca. From 2009 to 2012 he was a fellow of the graduate school 'Laboratory of Enlightenment' at the Friedrich-Schiller-University Jena, where he now works at the Ethics Centre and the Department of Applied Ethics. His research interests include human enhancement, especially the anthropological aspect, as well as political philosophy and phenomenology. He has published about Giorgio Agamben (2011) and is co-editor of *Politische Theorie und das Denken Martin Heideggers* (forthcoming).

**Benson A. Mulemi** is an anthropologist and senior lecturer in the Faculty of Arts and Social Sciences, at the Catholic University of Eastern Africa in Nairobi, Kenya. He is a visiting lecturer at the Institute of Anthropology, Gender and African Studies, University of Nairobi. He holds a PhD degree from the University of Amsterdam and a Master of Arts degree in Anthropology from the University of Nairobi, Kenya. His publications include contributions to *The Oxford Encyclopaedia of African Thought* and *African Folklore: An Encyclopaedia*.

**Anna G. Piotrowska** is associated with the Department of Theory and Anthropology of Music at the Institute of Musicology, Jagiellonian University in Kraków, Poland. Her research interests focus on the sociological and cultural aspects of musical life. She is the author of three books in Polish as well as numerous articles. Her book *Topos of Gypsy Music in European Culture* was awarded the honorary W. Felczak and H. Wereszycki Award by the Polish Historical Association in 2011. She was a Fulbright Fellow in Boston University, USA and in 2009 she was awarded the Moritz Csaky Preis at Austrian Academy of Sciences. She was also the recipient of a Mellon Fellowship awarded by Edinburgh University, UK.

**Christoph Rehmann-Sutter** teaches at the University of Lübeck in Germany and is a visiting professor at King's College, London. After a diploma in molecular biology, he studied philosophy and sociology. In 1996 he founded the Unit of Bioethics at the University of Basel, Switzerland, and he was president of the Swiss National Advisory Commission on Biomedical Ethics from 2001 to 2009. Research interests include the ethics of genetics, reproductive medicine, stem cell research, and end-of-life care. His books include *Leben beschreiben. Über Handlungszusammenhänge in der Biologie* (1996), *Zwischen den Molekülen. Beiträge zur Philosophie der Genetik* (2005), *Genes in Development: Re-Reading the Molecular Paradigm* (co-editor 2006), *Disclosure Dilemmas: Ethics of Genetic Prognosis after the 'Right to Know/Not to Know' Debate* (co-editor 2009).

**Christina Schües** is Professor of Anthropology and Ethics at the Institute for the History of Medicine and Science Studies, University of Lübeck. She is also Adjunct Professor of Philosophy at the Institute for Cultural Theory, Cultural Research and the Arts, Leuphana University, Lüneburg. She studied philosophy, political sciences, and literature in Hamburg and Philadelphia, USA. Her areas of research include anthropology, ethics, epistemology, phenomenology, political philosophy, and medical ethics. She especially aims at understanding the *conditio humana*, temporal dimensions of ethical conduct and of peace theories, concepts of the body, natality and generativity, and decision-making processes in medical and ethical contexts (e.g., in transplantation practices). Publications include *Philosophie des Geborenseins* (2008), *Time* *in Feminist Phenomenology* (ed. with Olkowski, D. and Fielding, H. (2011)), 'Menschliche Natur, glückliche Leben und zukünftige Ethik. Anthropologische und ethische Hinterfragungen', in *Verbesserte Körper und gutes Leben? Bioethik, Enhancement und die Disability Studies* (Eilers, M., Grüber, K. and Rehmann-Sutter, C. eds. 2012).

**Jackie Leach Scully** is Professor of Social Ethics and Bioethics, and Co-Director of the Policy, Ethics and Life Sciences Research Centre, Newcastle University, UK. She is also Honorary Senior Lecturer at the University of Sydney Medical School. After a first degree in Biochemistry and a doctorate in Molecular Biology she worked for some years in neuroscience research before becoming involved in bioethics. Her research interests include disability and identity, embodiment and disembodiment, feminist bioethics, reproductive and genetic bioethics, and the formation of individual and collective moral opinions. She is the author of *Disability Bioethics: Moral Bodies, Moral Difference* (2008), *Quaker Approaches to Moral Issues in Genetics* (2002), and co-editor of *Feminist Bioethics: At the Center, on the Margins* (2010).

**Tom Shakespeare** has researched and taught at the Universities of Cambridge, Sunderland, Leeds, and Newcastle. His books include *The Sexual Politics of Disability* (1996) and *Disability Rights and Wrongs* (2006). He is currently a technical officer in the Department of Violence and Injury Prevention and Disability at the World Health Organization, where he is one of the editors and authors of the *World Report on Disability*.

# 1
# Refocusing the Enhancement Debate

*Christoph Rehmann-Sutter, Miriam Eilers, and Katrin Grüber*

Superhumans (and the monsters in their shadow) have populated cultural imagery since the times of Mary Wollstonecraft Shelley and her Frankenstein. In Shelley's novel, the purpose of enhancement is improvement – 'what glory would attend the discovery, if I could banish disease from the human frame, and render man invulnerable to any but a violent death!' (Shelley 1993, p. 23) – but the results prove disastrous for those directly affected. The power to alter the human condition for the better by constructing better-abled bodies cannot be reduced to a simple matter of good or evil, but is itself deeply ambivalent. It may appear to be an irresistible, sometimes even morally tempting power, seemingly in the pursuit of good, but such a pursuit is also rife with hubris and produces many parallel disadvantages. This ambivalence is captured in the narratives of cultural imagery, where the beauties and the beasts, the heroes and the monsters, frequently appear as closely related, even as two personae of one and the same individual, one lurking just beneath the other.

The significance of enhancement, both in fiction and in reality, has not remained the same since Shelley's times, when the old experimental superhumans represented fictive attempts to conquer the roots of disease, evil, or death. To capture the intricacies of enhancement thinking and cyborg technologies, we can distinguish between different directions of 'improving' (see discussions in Hauskeller 2013): becoming smarter, making humans morally better, making them feel better, making them 'truly human', living longer, looking better, getting stronger, or compensating for natural frailty. We can also distinguish between currently possible interventions on the one hand – such as cosmetic augmentation of body shape, smart prostheses such as artificial limbs, mechanical exoskeletons, or implanted hearing aids – and more theoretical and futuristic ideas on the other – such as elimination of aging and death, deep genetic redesign of human bodies to something that is beyond human, or intelligent electronic brain implants to improve mental performance.

The debates about 'transhumanism' and future developments in enhancement are most frequently staged as a set of abstract moral questions (see Savulescu and Bostrom 2009), often referring to imaginary futuristic scenarios of what might be possible 'if and then' (Nordmann 2007, 31ff.). This discourse, as many of those who contribute to it themselves have complained, lacks concreteness. We believe that the perspective of disability will contribute towards substantiating the enhancement debate.

## Disability studies

Even though the question seems to be an obvious one, it has rarely been raised in academic bioethics debates on enhancement biotechnologies: how does enhancement relate to disability? At first glance, the relationship might seem simple: enhancement is the gain of function, disability is the loss of function. One relates to something above normal functioning and the other below, like mirror images with the level of normality as their axis of reflection. A closer look, however, reveals a tremendously rich and complex relationship, whose exploration proves fruitful in discovering what enhancement is really about.

Disability studies, an interdisciplinary field of anthropological, social, philosophical, and political research, have shown that the common view of non-disabled people of disability as a loss of function is both far too narrow and far too discriminatory. If we see disability as implying a life of missed possibilities and opportunities, as a 'harmed condition' (Harris 1993, 180) or a 'physical or mental condition we have a strong rational preference not to be in,' (Ibid.) as John Harris has described it, (see Chapter 9, this volume) we are adopting the general perspective that has been described as the 'medical model' of disability: 'From the medical point of view, people are disabled when they are less functionally proficient than is commonplace for humans, and when their dysfunction is associated with a biological anomaly' (Satz and Silvers 2000, p. 173). Medicine, accordingly, aims to reduce, preferably to cure, such dysfunction. According to disability studies, operating within the medical model of disability disregards crucial contextual factors, and sometimes, rather than easing the physical or mental impairments of those classified as disabled, actually creates further problems.

The founding credo of disability studies has been the 'social model' of disability, which shifts the attention away from the physical or mental impairment to the living conditions and societal situation in which an impairment becomes a disability and manifests itself as a problem. The approach started as a 'materialist' one in the sense that it looks for the social causes of all harmful or negative aspects of disability. A disadvantage of this approach is that it tends to overlook the effects of impairment on the individual experience of disability and therefore cannot really explore the differences between different impairment groups.

The most recent versions of the social model of disability consider that the experience of disability is a composite of factors that emerge within a wide context. However, there are also intrinsic factors, such as the type and severity of impairment, the attitude of the person towards this impairment, and her or his personal characteristics and possibilities. Furthermore, there are extrinsic factors, such as the attitudes and reactions of others, the presence of a supporting or disabling environment, and further cultural, social, and economic issues. Proponents of the social model approach in disability studies have experienced some difficulties acknowledging this ambivalence in taking into account the condition of the body while also working on the grounds that well-being depends on social circumstances and society's capacity to recognize all of the specific needs of people with variant bodies beyond simply the internal/external condition of the body (Shakespeare 2006, pp. 29ff.; Williams 2001). In one consensus perspective, disability can be viewed as 'a dynamic interaction between health conditions and contextual factors, both personal and environmental' (World Health Organization and The World Bank 2011, p. 4).

The social model approach indicates that medical improvements do not always have a positive effect on people with a disability; rather, their effects may be multiple and as dependent on social and cultural factors as the disability itself. It considers the limitations of the assumptions made by those with 'normal' bodies of what life is (including notions of a 'good life') for people with variant embodiment and of the biopolitics of human difference. This way of thinking is sensitive to the social conditions necessary for leading a good life, including the importance of variant bodies (Blume 2012; Scully 2008; Shakespeare 2006; Vehmas 2012). The perspective of disability studies might also be necessary, therefore, to bring the enhancement debate onto more reliable ground.

Tom Shakespeare (2013), in capturing the state of current research in disability studies, distinguishes different directions or approaches to disability. Apart from the 'materialist' approach there is a direction called the 'cultural' approach, in the sense that it looks for the social and cultural construction of the dichotomies that underlie the experience of disability and normality. Disability itself becomes a social construct that needs to be deconstructed, using the analytical instruments of poststructuralism or critical cultural studies. Another direction that also tries to meet this difficulty can be called the 'realist' approach (or better, 'critical realist' or 'post-social-model'), in the sense that it is primarily interested in the empirical investigation of the life conditions and the needs of differently impaired and disabled people in different places, countries, and situations. We believe that all of these different approaches to disability might have something to contribute with regard to the topics discussed in this book. The chapters will draw in different ways on the concepts from these three approaches to disability, as well as on existing knowledge about the phenomena and social processes of disability.

Our working hypothesis is that connecting the two fields of enhancement bioethics and disability studies brings both forward and infuses each with a new set of questions. Just as the enhancement debate lacks the perspective of disability, disability studies lack the question of enhancement and technology development. The contributors to this volume have thus investigated the intimate, ambiguous, and in many ways significant relationship between good lives and (better) bodies, and the role that enhancement technologies could and should play in this relationship. It is an interdisciplinary book that combines philosophical, anthropological, and sociological disciplines with cultural studies. It aims to invite more grounded discussions about enhancement, which set aside the counterproductive and reductive division between dis- and plus-ability research.

The authors assembled in this book do not all share the same approach to disability, but they are all skeptical about the medical model of disability and about biomedical practices that reduce the experience of people with disabilities to their impairments. We acknowledge the importance of medical progress for disabled people, but this in no way excludes the possibility of skepticism regarding the medical model in terms of explaining disability. Medical-technical developments in many fields have improved the lives of people with a disability, and many experience severe problems if they do not have access to the health care system (Grüber 2012; WHO and WB 2011). The medical and ethical issues connected to strategies to find appropriate solutions for the problems that people with disabilities face are, however, not the topic of our book. Rather, the central issues in this volume are questions around the concepts of normality and normativity, which contribute both to the enhancement debate and to disability studies.

Where does the idea of normality, which is crucial for the therapy–enhancement distinction, come from? How relevant is the concept of normality? What is the hidden normativity within the concept of normality? And what kind of normality is relevant? Is it

- the statistically average functionality in all species members;
- species-typical normality (whatever the word 'typical' means here – for example, an ideal, as in the sociological term 'ideal-typical', or a necessary minimum just fit for survival of the species); or
- an individually experienced state of accustomed embodiment?

It is important to look deeper into these concepts in order to obtain a clearer picture of the possibilities of enhancement.

## The biopolitics of a debate

There are several reasons why the enhancement debate excludes the perspective of disability and disability studies. For one, the perspective of disabled people is generally underrepresented in the various debates within bioethics (Scully 2008). Another concrete reason might be the widely shared assumption that enhancement is conceptually distinct from repair, therapy, or prosthetics. This distinction is defended by referring to the idea of normality: enhancement is what brings the human body beyond normal functioning, whereas repair and therapy bring it back to normal. In this sense, prostheses or disability aids should imitate a normal embodied function, replacing the natural limb or sense with an artificial device. Even though they may enhance the shape and function of an individual's impaired body, they are therefore not enhancements.

The ethical issues of medically assisting persons with disabilities in their lives or the ethics of medically eliminating disabilities are separate from those ethical issues that surround interventions that go 'beyond therapy' (President's Council 2003). This framing of biotechnology as having the potential to repair or enhance is the standard 'order of things'. It implies – like all order – politics (Foucault 1974). If it is taken for granted, disability marks the space underneath the table at which 'normals' take their seat. Normality is, moreover, constructed as a dividing zone, from which some forms of embodiment deviate – either towards the negative or the positive.

The discourse on the social and ethical implications of human enhancement technologies focuses on the issue of moral permissibility. Is it morally permissible to enhance in a specific way? Or are enhancing interventions intrinsically wrong; should they be banned by law? The objects of controversy are frequently hypothetical endeavors, such as advanced germ-line technologies for creating new generations of children with a longer health span, creating increased 'moral strength' or additional senses through the use of imaginary nano-bio-cogno devices to be implanted into the brain, or even the idea of switching human life forms to a digital form of existence through uploading minds onto computers. In a moral and also regulatory perspective, the main question of interest here is: should somebody who has the desire or interest to do such things be allowed to do them or not? This is a difficult question, for the simple reason that we do not know enough in order to address it properly; this is why enhancement is a challenge that calls for 'postconventional' ideas (Shildrick 2005). Enhancement is, insofar as it puts the question of the *conditio humana* at stake, not only a futuristic discourse. It also points to the present, since it is already modifying the current academic and public dialogue about the body, as well as taking ethics to the past. How, therefore, can concepts from modernity fit postmodern (such as enhancement) discussions (Shildrick 2005, pp. 3–4.)?

How can ethics – which far too often takes 'what it means to be human' for granted and has a strong tendency to rely on modern templates when imagining new (future) knowledge (Ibid.) – take a stance on enhancement? One answer would be to admit the unfamiliarity of the consequences of such knowledge and pursue the link between enhancement and disability, as both an example of thought and a training of thinking. In this sense, our book contributes towards reconfiguring the ethical framework of the enhancement debate.

Michael Hauskeller pulls attention in a slightly different direction when he observes that enhancement, in a strict logical sense, is rather impossible: 'we lack any clear idea of what it would actually consist in without being aware of that lack' (Hauskeller 2013, p. 186). The lack of awareness of this lack of insight is dangerous because it is not recognized. The enhancement projects we are talking about are planned within an instrumental rationality. Such instrumental rationality, however, has its limits, because it is not clear whether the improvement (according to whatever scale) will be a real betterment overall for the persons affected and for their social relationships, or whether it will be at least ambivalent, or possibly even prove to be harmful. What seems to be good in certain respects might be bad in others. What is clear is that enhancement would, in a certain respect, imply 'more than good': we should demand an improvement of the human condition or human well-being.

What would the relevant understandings of 'morality' and 'the good' be in order to decide upon such questions? On what grounds could someone be denied access to enhancement technologies (if available), and on what grounds could the application of such technologies be ethically legitimized?

It would of course be better to begin to explore this murky terrain of upcoming biotechnology early and assess its potential effects well before the fact, that is, upstream in the flow of biotechnological innovation. However, it is not so clear how to frame the ethical questions at this point, when many things are still so vague and hypothetical (Rehmann-Sutter and Scully 2010). Even though some scientists and scholars do argue that at least some enhancement developments are already quite realistic, or will be soon, it would still be early enough in terms of such developments to say 'no' if necessary. What is perhaps more important is that there is still room to understand and frame such developments in appropriate ways, since the technological and therapeutic imperatives have not yet been reshaped and the persuading 'normative power of facticity' ('use them because they are available and others use them too!') is not yet overwhelming.

## A phantom?

With good reason, many of the visions discussed in the enhancement discourse can be seen as unrealistic or purely hypothetical. In a review article, pharmaco-psychologist Boris Quednow (2010) called the discussion on cognitive enhancement drugs a 'phantom debate' because it assumes that a technology 'that will probably never materialize' (Quednow 2010, p. 156) is in fact realistic. He states: 'The assumption that, in the near future, we will have access to compounds that are not only effective cognitive enhancers, but also safe and well tolerated and therefore suitable to be taken by everybody' (p. 154) is unlikely to be true. The same could be said about other enhancement visions. Gregory Stock's 'genetic molecule' that one could 'safely add to an embryo and thereby give your future child extra decades of healthy life' (Stock 2002, p. 78) is intentionally kept vague. Molecular biologists are careful not to overstate the progress of genomics (Lander 2011), as we are currently far from seeing realistic genetic augmentation, and perhaps never will arrive at those radical enhancements that transhumanists ponder over.

This gap between what is dreamed of, feared, and debated and what is actually possible means that we need to reflect on the phenomenon of enhancement and its associated debates, to view it as a cultural phenomenon. The discipline of science and technology studies can contribute here, as it enables us to see technologies not as a given but as contingent and contextual. Through this lens, we can take a closer look at the consequences, at the 'decisions, the trade-offs, the evidences and their interpretation within certain political strategies' (Blume 2012, p. 352). Such an approach allows for the possibility to actually shape the developments and technologies, instead of simply observing and documenting them.

As pointed out above, the enhancement debate has two sides: one which focuses on the applications already in practice, and the other which reflects on future visions. By drawing attention to this distinction, and to the experiences with enhancing biotechnologies that are already available, we hope to counteract a tendency in the debate to stay on a (sometimes astoundingly) high level of abstraction and generality regarding the question of duties, aims, permissibility, and the illicit. Even if unrealistic, such visions can be seen, as for instance Peter Wehling (2011) has suggested, from a Foucauldian perspective on biopower, as one part of a larger biopolitical process of technification. Such visions have the potential, if realized, to dramatically change the climate in the public sphere; for example, by discursively constructing the 'natural' human body as 'imperfect' compared to those that are enhanced. The debate might be based on unrealistic plans or expectations, but it may nonetheless have a real effect on what people see as the limits of the human condition. Furthermore, it has the potential to delimit moral perception. Even taking into account the sometimes exaggerated leaps in technology that enhancement proponents assume, the debate and its framing still prepare people to see the human body and its basic structures as a legitimate object of (at least reasonable) biotechnological improvements; whatever 'reasonable' means in any given time and place.

We, the editors and authors of this book, do not want to fall into such assumptive traps, nor do we intend to set them. One advantage of adopting such a cautious attitude is that it helps in terms of being wary of the common conclusion that we are morally obligated to use genetic technologies to produce 'the best children possible' and to 'improve ourselves' (Savulescu and Kahane 2009; Sparrow 2011). Such an obligation seems to be derived from moral principles of beneficence and autonomy, meaning from our duty to promote the well-being of children without reducing their freedom of choice. This value-driven approach is based on questionable assumptions. Firstly, it assumes that if something is an enhancement of the human body, in certain respects it will indeed promote the well-being of those who live with this kind of body. To equate enhancement (of a desired function or feature) with a higher state of well-being in this way is unlikely to be true, since in some cases (such as muscle strength or longevity), this function is dependent on other conditions in order to be a contribution to well-being, and in other cases (such as memory), the function must be well balanced with an opposite (such as the capacity to forget). Secondly, it assumes that we know enough about future humans' life plans to be in the position to decide about the necessary conditions for their well-being. If only of these assumptions questionable, the conclusion that we have a moral obligation to enhance our children because we have an obligation to improve their well-being is called into question.

## From visions to real life

One reason why the enhancement debate is abstract might be the unavoidable lack of evidence with regard to the contextualized real life impacts of futuristic biomedical if-and-then scenarios (Nordmann 2007). When we enlarge the focus and concentrate on the 'now', we can detect some medical interventions and devices that could, depending on the concept of normality one is using, also be viewed as enhancement technologies: for instance, cochlear implants and high-tech prostheses. There is no sharp line between restoration and enhancement because the device never provides a one-to-one replacement of a lost body part, and the selectively added functional abilities can rarely be measured against the ones that were replaced. Experiences of the real life impact and socio-political implications of medical devices are available and, to an extent, systematically studied. These experiences are, however, frequently expatriated from the bioethical enhancement debate because, according to the dominant view, prostheses 'restore' but do not 'enhance'. But what is seen as 'just restoring' versus 'enhancing' depends on the definition of the measurement that defines these concepts; namely, the concept of 'normality'. For people who live with certain disabilities or chronic illnesses, their 'normal' is different from the generally assumed 'normal' that is proposed by medicine, or from the 'normal' that is seen by population biologists or public health theorists as 'species-typical functioning'. There is a 'felt status of normality' (Scully 2008, p. 100) for people with variant embodiment, dependent on the nature of this variation. Therefore, the systematic expatriation of evidence from the lives of people with a disability or chronic illness, using the argument that there is a clear divide between restoration and enhancement, rests on assumptions that in fact firmly belong within – and should also be the subject of – the debate.

Much can be learnt from case studies, such as those provided by two contributions in this book (chapters 5 and 6). Furthermore, there are documented experiences with non-genetic, pharmacological, and surgical body improvements – such as the augmentation of the eyelids, nose, breasts, vagina, and penis, or liposuction – that are relevant in order to see how technology works in everyday life, how needs are generated, how bodies are conventionalized, and how the corresponding risks are made acceptable. Such technologies and their associated experiences represent individualized solutions for social problems; for instance, when cosmetic surgery, if used to meet or even transpose social norms of beauty and sexual attractiveness, provides better chances for success. Research in science and technology studies can help to understand these mechanisms.

## Ambivalence of biomedical interventions

The abstraction of the discourse has even more direct consequences. It ignores the rich wealth of people's experiences of the ambivalent nature of many biomedical interventions, which always have both up- and downsides. Every medical intervention has effects and side effects, and while they might indeed contribute to well-being, they might also have costs for those affected, or for others. Furthermore, every cure is selective, choosing from among a plurality of possibilities for betterment; some symptoms or capacities and not others. Prosthetic devices, for instance, do not restore the full range of functions of a lost limb, but highlight some functions and exclude others. The leg that has been replaced by a prosthesis can walk, run, move, and give the person an appearance, but it could potentially also feel, touch, or be desired. Prostheses are only functional equivalents in some aspects, while they defer others. They may even bring one function of the limb to a level beyond its 'natural' functionality, or add functions that were not there before. Prostheses are therefore inherently selective (Eilers 2012).

We propose to ground the abstract enhancement debate in anthropological reflections on technology and disability experiences. Using this approach, we aim to better develop and refine our understandings of the ambivalences of biomedical interventions, since to acknowledge ambivalence is not to reject a development but to consider it in depth, which will in turn lay the ground for better judgment.

## Recognition of the body

A second reason why the current bioethical enhancement debate is abstract might be the preferred argumentative style in mainstream bioethics, which is frequently preoccupied with general statements about attitudes, norms, values, principles, and the reasons behind them, rather than with embodied everyday experiences, contextualized moral perceptions, and the ethical reflexivity of ordinary people. Mainstream bioethics, as Margaret Shildrick (2005, p. 2) has observed, is 'out of touch with bodies themselves, in the phenomenological sense'. The self, with its desires and its wishes to be better than given, is 'always intricately interwoven with the fabric of the body' (Ibid.).

What the enhancement debate can add to moral philosophy is the issue of embodiment. A good life is necessarily corporeal, and '[s]ince we are fundamentally corporeal beings, our existence is characterized by various physical needs and dependencies throughout our lives' (Vehmas 2012, p. 307). This remains true even if some transhumanists propose leaving the biological body behind through the uploading of the mind onto a computer (Bostrom 2003, 17; see Chapter 11, this volume, for a critical evaluation of this vision). Assuming that such a vision would be achievable, this inhabited computer would thus become the person's body. Disability studies contribute the perspective that differences in embodiment, under certain circumstances and together with social conditions, become crucial and indeed affect the sphere of the good and desirable in human life. Thinking through the variant human body captures questions about the good life in a different way, starting with a thorough phenomenology (Rothfield 2005) that acknowledges difference (see Scully 2008; Chapter 2, this volume).

## The good life, the better life, and well-being

One topic directly connects the fields of bioethics and disability studies: the complex idea of the relationship between well-being or 'human flourishing' and the body. While an enhanced body is claimed to have positive impacts on the well-being of a person, a disabled body is seen as a cause for reduced well-being. But very often this is simply an assumption without an evidential basis. Well-being, if it is scrutinized in detail on the basis of diverse and complex experiences, and analyzed in its philosophical depth, is far from such a 'narrow metric' (Sparrow 2011, p. 36). Furthermore, '[e]ven so-called non-disabled people achieve forms of well-being in degrees that vary greatly. Impairment is merely one factor among many other things that may affect one's chances to pursue different dimensions of being well' (Vehmas 2012, p. 303).

Well-being and human flourishing directly relate to a key topic in practical philosophy: the 'good life' (Nussbaum and Sen 1993; Schües and Rehmann-Sutter 2013; Chapter 3, this volume). Well-being and flourishing cannot be understood, questioned, or explained without reference to ideas of what is really worthy of being desired in life. Well-being thus has a normative ethical component as well as an empirical or experiential one. In both respects, well-being is a richer concept than happiness, which could simplistically be understood as an emotional state, perhaps something like a warm and bright inner feeling. Ethical reflection on the good life does not exclude feelings of happiness, but essentially covers the full range of human capabilities, practices, and functions, by, for instance, including the capacity to cope with unhappiness, the positive sides of melancholy, and the circumstances that lead to it.

Ethical reflections on the good life date back to Aristotle (Nussbaum 1994; Wolf 1994; Wolf 1999), and are a reflection on the ultimate, desirable objectives in human life, that is the aims and goals that make life worth living. Every concrete project of human enhancement (if the term is used deliberately, meaning a path to the better) must therefore encapsulate an idea of 'the good'. These ideas need to be unwrapped, brought out into the open, and discussed from a wide range of perspectives, as the authors of several chapters of this book do. They work towards a more balanced view, one that includes ambivalent elements associated with the idea of the allegedly good.

To understand enhancement for the purpose of a good life and well-being, a research perspective that takes into account all approaches to disability within current disability studies is required. Both the disabilities *and* abilities inherent and implied in both the 'normal' state of embodiment (which should be enhanced) and the 'enhanced' state of embodiment (which might be ambivalent) need to be investigated. They need to be observed from the point of view of their social etiologies (strong social model of disability), their discursive construction (cultural model), and their impact on life conditions (post-social-model). The disabilities as well as the abilities of normal and enhanced states need to be critically assessed from these different perspectives. What we propose adding to disability studies with this volume is therefore a set of research topics rarely treated under the label of disability: the abilities and disabilities that are involved in enhancement projects and practices. They appear both before and after the enhancing interventions: before the intervention when a state is seen as in need of improvement, and after the intervention when the resulting life form is not uncontroversially good but rather a mix of abilities and disabilities, old and new.

## Beyond the normal

Enhancement is unavoidably a value-laden term. As we have said, it is very often seen in contrast to 'therapy', that is, the reinstatement of a normal state of functioning after disruptive incidents of disease. This provokes several questions. For starters, what is a normal state of functioning (especially from a disability perspective)? The second question emerges from the fact that enhancement, according to the common use of the term, goes beyond the normal state when it claims that the state above normal is not just different but also better; that 'more' equals 'better'. Looking at such an assumption critically, we need to ask on which scale this 'better' is evaluated, and furthermore, can the good in a human sense be scaled at all?

Discussion of these topics among the contributing authors, who come from different disciplinary backgrounds (philosophy, bioethics, medicine, sociology, disability studies, and medical anthropology), developed from a basis of shared mistrust of the dichotomous definition of enhancement versus therapy, in both descriptive and prescriptive terms. It is not entirely clear what 'normal functioning' actually is; is it the species average, the functioning of the best (least disabled, healthiest, strongest...) naturally occurring species members, or is it an individual benchmark indicating what everyone sees as normal for her – or himself, regardless of whether it fits with any group or species norms? The contributors' hesitations were also bolstered by the experiences of people with disabilities who are acutely alert to the tendency in many societies to use definitions of normality for discriminatory purposes: whoever does not fulfill the norm might not be recognized as equal. If a species norm is used as a normative criterion to distinguish morally acceptable from unacceptable interventions, discriminatory use of the normal/abnormal frame could thus be reinforced.

Enhancement is a value-laden and unclear term; but what would be more neutral and clear? We have explored different alternatives, the most useful of which was simply speaking of 'altering' technologies: biotechnologies that are not intended to cure or help but to change a given state of the body. The advantage of deconstructing enhancement is that it brings the question of the presupposed interests involved out into the open. If the intention is to change something, we should always ask for what purpose this is done. Is it a purpose the person identifies with? Is it an external purpose, perhaps internalized by the individual via suggestive cultural frameworks? What would be the unintended consequences within a society where people with disabilities live together not only with 'normals' but also with 'supernormals'? What is implied in the transhumanist approach where everybody who is not enhanced is by default disabled (see Chapter 9, this volume)? Visions of enhancement are closely linked to particular understandings of human society, which also need to be scrutinized.

Finally, norms of regulation can both govern and discriminate. Good governance of enhancement should avoid a blanket ban simply for the reason that it takes humans beyond the 'normal' or the 'natural'. Such a stance would transform statistical normal functioning or normal states into a moral criterion to use when distinguishing between permissible and impermissible interventions (Scully and Rehmann-Sutter 2001). Such a move would increase the tendency of societies to first invent normal states and then declare them 'natural' or 'right', and the deviations from them as 'abnormal', 'unnatural', and 'wrong'.

## Conclusion: Why enhancement bioethics needs disability studies

In bioethics, disability can function as a trigger that opens up a rich socio-cultural framework. Within this framework, we can discuss enhancement on more levels and perceive its actual and potential reality in a broader context. Beyond the search for the appropriate moral distinctions and governance of enhancement biotechnologies, and beyond the quarrel about the right ethical theories that might help to find them, there is something deeper. Enhancement is also about renegotiating the significance of variant embodiment for a good life. It is about identity, about culture versus nature, and about the role that 'naturalness' plays in conceptions of subjectivity and identity. It is about the coherence of the human species and the defining features of social groups. It is about the ethos of corporeality, the sense of completeness and integrity of embodiment. It is about what wholeness means, and for whom.

More concretely, there are, in our view, several reasons why knowledge about disability is important for the enhancement debate. The experience of disability, limitation, and illness can help put the question about the good life, which is raised by enhancement projects, on more solid and experiential ground. The issues that are raised in the debate can thus be more concretely understood. In particular, the experience of disability can prevent bioethics from inadvertently accepting the 'medical model' of functional improvement. We cannot really understand what an actual improvement of the body is without taking into account the knowledge and experience of variant embodiment of people with a disability.

If the experience of disability cannot be comprehensively understood without acknowledging all the intrinsic and extrinsic or contextual factors, then it is unlikely that the experience of enhancement can be fully understood without acknowledging all of these factors. In other words, it is insufficient to look only at the functional aspects, capacities, or features of the body that should be 'improved'. Contexts can turn the experienced value of a functional improvement into its opposite, just as they can change the experienced value of a functional impairment. The biomedical improvement of a function may therefore be insufficient to enable a good or better life, since there are structural and contextual factors that might be more important for building actual capabilities.

We believe that disability studies can prevent bioethics from relying too much on abstract thinking and can encourage cooperation with and integration of the experiences of those with varied embodiment. Furthermore, the aim of disability studies is not to develop an abstract conception of disability from an external, objective perspective. Rather, it is important to start from the real experiences of people who have disabilities and live with them. Improvement of life needs to be an improvement of the lived life, in the same sense that the concept of disability should necessarily be based on the experiences of people with disabilities.

This has an immediate application. The experiences of a person with a disability of medical interventions aimed at improving life demonstrate that side effects – both corporeal and social – are unavoidable. One of the most well documented cases is the cochlear implant (which has been investigated by Blume 2009; cf. Chapter 5 in this volume). The scientists and clinicians who believed that they were providing a cure for the loss of hearing by purely technical means did not realize that not all Deaf people actually experience deafness as a loss of function of the ear, as something that needs to be cured. There are many Deaf people, especially those who were born deaf, who do not see themselves as disabled but rather as a member of a minority group with a rich culture ('Deaf culture') and its own language. In the early days of the technology, Deaf persons did not wear cochlear implants, but over time the situation has become more complex and a more nuanced view among Deaf people has emerged. There are 'culturally Deaf people who are happy to use hearing aids or cochlear implants (CIs) and those who reject them as unnecessary or even as threats to the integrity of their identification as Deaf' (Scully 2012, p. 113). Similar developments might happen if technological improvements are directed to other functional capacities, especially if only a medical model of the corresponding function or dysfunction is recognized.

Disability studies make bioethics more alert to discriminatory effects. Variant or deviant embodiment (differing from whatever standard), as experience demonstrates, is frequently connected with discrimination (see the *World Report on Disability* of the World Health Organization and The World Bank 2011). It is not such an implausible extrapolation that the 'victims' or 'beneficiaries' of enhancement could also be affected. There are many points where prejudice, discrimination, and social injustice may emerge, even if it is unclear on which side these conflicts will start (see Chapter 9, this volume). Discrimination can run in two directions: from the improved to the non-improved, or from the non-improved to the improved. Those who are improved will not necessarily be seen by the non-improved as peers. One group could have an advantage over the other; the improved, for instance, could possess a capacity that maybe some non-improved people would also like to have. But the improved person could in certain circumstances – in school, for example – still be dependent on the social recognition of non-improved persons. Correspondingly, enhancement could also aggravate discrimination of those who are already marginalized, such as people with disabilities. Furthermore, even though it might be desirable, having a stronger or healthier body would not automatically mean being more empathetic with those with fewer abilities (see Chapter 2, this volume).

Those privileged people with an allegedly stronger body would be those whose life or career is a bit easier in a certain respect. Those able to enjoy a longer life than others could benefit longer from social insurances, for example. From the perspective of social justice and injustice, the distribution of privileges would thus also need to be evaluated. If, in the future, human beings are created that are stronger, faster, prettier, and longer living, the difference between them and those who are less strong, less fast, less pretty, and who live shorter lives will be wider than at present. Perhaps most importantly, it is crucial to acknowledge that this increased gap between different embodiments would be human-made. This is one example that demonstrates the importance of bringing together dis- and plus-ability research.

We hope to illustrate in this volume why it is necessary to enlarge the focus of the moral debate – perhaps sometimes even bracketing the moral question, suspending it for a moment – in order to better understand the complex relationship between enhancement and disability, before making any final judgments.

## The structure of the book

The book is divided into three parts, which address the human enhancement debate from different working perspectives.

Debates about enhancing the human often focus on the brain as the main subject (neuro-enhancement). In contrast to this, the first part of this book places the whole human body into the center of interest. The first two contributions introduce the body as the entity that is a necessary biological condition for the human lifespan, but which is at the same time susceptible to cultural change. The authors elaborate on the historical meanings of the body and its political implications, as well as the norms to which the body has been – and still is – subjected.

The authors of the second part take these findings and, on the basis of case studies, work on the question of what variant embodiment means for individuals. We could say that they start from Wittgenstein's statement that 'If someone says, "I have a body", he can be asked: "Who is speaking here with this mouth?"' (Wittgenstein 1970, §244). The authors in this section tell of deafness, altered sexuality, and old age; of dealing with cognitive impairments like depression; and they reflect on the consequences of invasive cancer therapy. In doing so, they first clarify what alteration means in an individual case and how medical interventions affect a diversity of embodiments. The authors then refer back to the individuals and consider what changing the mind and the body means for less mind-centered but nevertheless crucial concepts such as authenticity and vulnerability. The texts try to clarify how biomedical interventions lead to improvement, impairment, or both; sometimes accepting one in favor of the other.

In the third part, the contributors converge on the implications of the enhancement debate. As said above, it is questionable whether all of the visions of human enhancement that are currently being debated will even be possible or actually implemented technically (and socially). From this point of view, the debate reveals more about our present cultural dreams and hopes than it indicates the real future. The debate is therefore twofold in more than one regard: it comprises present and future, possibilities and dreams, and turns on our current hopes and wishes about corporeality. In this final part, the authors have searched for cultural deployments of human enhancement and found them in sources ranging from old myths to contemporary science fiction films. Others discuss arguments from philosophical phenomenology, which favor the fallible body and confront the futuristic body with the unimproved one.

We hope that the book, which is inspired by multiple approaches, will make a contribution to the enhancement debate. Even if one remains (as we do) skeptical about human enhancement biotechnologies, this does not imply that one is necessarily a bio-conservative. Rather, we believe that an interdisciplinary and culturally informed bioethics is necessary in order to explore the grey zones.

## Acknowledgements

We owe special thanks to several persons who have made this book possible. It is one of the outcomes of the one-week workshop entitled 'Good life better – anthropological, sociological and philosophical dimensions of enhancement', which we the editors organized in October 2010 in Lübeck, Germany. A different set of papers has been published in German (Eilers et al. 2012). The workshop and the publication project was funded by the German Federal Ministry of Education and Research (Bundesministerium für Bildung und Forschung, BMBF) under the program 'Klausurwochen auf dem Gebiet der ethischen, rechtlichen und sozialen Aspekte der modernen Lebenswissenschaften', and was hosted by the Institute for the History of Medicine and Science Studies of the University of Lübeck and the Institut Mensch, Ethik und Wissenschaft (IMEW GmbH), Berlin. The idea to connect the enhancement discourse with disability studies traces back to Katrin Grüber, who chaired a conference exploring this topic in 2009 in Linköping, Sweden, which was funded by the European Science Foundation. The editors give thanks to the BMBF for a pleasant and uncomplicated collaboration, notably to Simone Mistry and Matthias von Witsch at the German Aerospace Center (Deutsches Zentrum für Luft- und Raumfahrt). We thank Anja Bracke, Kathrin Hoffmann, Angela Mötsch, and Evelyn Österreich (Institut für Medizingeschichte und Wissenschaftsforschung der Universität zu Lübeck) for their support with planning and conducting the conference and this book. Zoe Goldstein re-read all chapters and made very helpful suggestions to improve the texts. Monica Buckland and Jackie Leach Scully translated one chapter from German to English and revised several other chapters linguistically. Special thanks to Tom Shakespeare who enriched this book by contributing a preface. Finally, our very warm thanks go to all participants of the Lübeck workshop.

## References

Blume, S. (2009): *The Artificial Ear: Cochlear Implants and the Culture of Deafness*. New Brunswick NJ: Rutgers University Press.

Blume, S. (2012): Science and technology and disability? In: Watson, N., Roulstone, A. and Thomas, C. (eds.) *Routledge Handbook on Disability Studies*. London: Routledge, pp. 348–359.

Bostrom, N. (2003) *The Transhumanist FAQ. A General Introduction*, version 2.1., World Transhumanist Association, http://www.transhumanism.org/resources/FAQv21.pdf.

Eilers, M. (2012): 'Fünfundzwanzigstündiger Arbeitstag – denn 'ne Prothese wird nie müde.' Normative und selektive Implikationen der Prothetik nach dem Ersten Weltkrieg. In: Eilers et al. (eds.) (2012), pp. 165–180.

Eilers, M., Grüber, K. and Rehmann-Sutter, C. (eds.) (2012): *Verbesserte Körper – gutes Leben? Bioethik, Enhancement und die Disability-Studies*. Frankfurt/Main: Lang.

Foucault, M. (1974): *Die Ordnung des Diskurses*. München: Hanser.

Grüber, K. (2012): Bedingungen für ein gutes Leben mit Behinderung. In: Eilers et al. (eds.) (2012), pp. 89–105.

Hauskeller, M. (2013): *Better Humans? Understanding the Enhancement Project*. London: Acumen.

Lander, E. (2011): Initial impact of the sequencing of the human genome. In: *Nature*, 470, pp. 187–197.

Nordmann, A. (2007): If and then: A critique of speculative nanoethics. In: *Nanoethics*, 1, pp. 31–46.

Nussbaum, M. C. (1994): *The Therapy of Desire. Theory and Practice in Hellenistic Ethics*. Princeton: Princeton University Press.

Nussbaum, M. C. and Sen, A. (eds.) (1993): *The Quality of Life*. Oxford: Clarendon Press.

President's Council on Bioethics (2003): *Beyond Therapy: Biotechnology and the Pursuit of Happiness*. New York: Dana Press.

Quednow, B. (2010): Ethics of neuroenhancement: A phantom debate. In: *BioSocieties*, 5, pp. 153–156.

Rehmann-Sutter, C. and Scully, J. L. (2001): When Norms Normalize. The Case of Genetic Enhancement. In: *Human Gene Therapy*, 12, pp. 87–96.

Rehmann-Sutter, C. and Scully, J. L. (2010): Which ethics for (of) the nanotechnologies? In: Kaiser, M., Kurath, M., Maasen, S. and

Rehmann-Sutter, C. (eds.) *Governing Future Technologies. Nanotechnologies and the Rise of an Assessment Regime*. Berlin: Springer, pp. 233–252.
Rothfield, P. (2005): Attending to difference: Phenomenology and bioethics. In: Shildrick, M. and Mykitiuk, R. (eds.) *Ethics of the Body. Postconventional Challenges*. Cambridge: MIT Press, pp. 29–48.
Satz, A. and Silvers, A. (2000): Disability and biotechnology. In: Murray, T. (ed.) *Encyclopaedia of Ethical, Legal, and Policy Issues in Biotechnology*. New York: Wiley-Interscience, pp. 173–187.
Savulescu, J. and Bostrom, N. (eds.) (2009): *Human Enhancement*. Oxford: Oxford University Press.
Savulescu, J. and Kahane, G. (2009): The moral obligation to create children with the best chance of the best life. In: *Bioethics* 23: pp. 274–290.
Schües, C. and Rehmann-Sutter, C. (2013): The well- and unwell-being of a child. In: *Topoi* 32: pp. 197–205.
Scully, J. L. (2008): *Disability Bioethics. Moral Bodies, Moral Difference*. Lanham: Rowman & Littlefield.
Scully, J. L. (2012): Deaf identities in disability studies, with or without us? In: Nick Watson, A. R. and Thomas, C. (eds.) *Routledge Handbook on Disability Studies*. London: Routledge, pp. 109–121.
Shakespeare, T. (2006): *Disability Rights and Wrongs*. London: Routledge.
Shakespeare, T (2013): *Disability Rights and Wrongs Revisited*. London: Routledge.
Shelley, M. (1993): Frankenstein. Or: The modern Prometheus. In: Butler, M. (ed.) *The 1818 Text*. Oxford: Oxford University Press.
Shildrick, M. (2005): Beyond the body of bioethics. Challenging the conventions. In: Shildrick, M. and Mykitiuk, R. (eds.) *Ethics of the Body. Postconventional Challenges*. Cambridge: MIT Press, pp. 1–26.
Sparrow, R. (2011): A not-so-new eugenics. Harris and Savulescu on human enhancement. In: *The Hastings Center Report*, 41(1), pp. 32–42.
Stock, Gregory (2002): *Redesigning Humans. Our Inevitable Genetic Future*. New York: Houghton Mifflin.
Vehmas, S. (2012): What can philosophy tell us about disability. In: Watson, N., Roulstone, A. and Thomas, C. (eds.) *Routledge Handbook on Disability Studies*. London: Routledge, pp. 298–309.
Wehling, P. (2011): Biopolitik in Zeiten des Enhancements. Von der Normalisierung zur Optimierung. In: Dickel, S., Franzen, M. and Kehl, C. (eds.) *Herausforderung Biomedizin. Gesellschaftliche Deutung und soziale Praxis*. Bielefeld: Transcript, pp. 233–250.
Williams, G. (2001): Theorizing disability. In: Albrecht, G. L., Seelman, K. D. and Bury, M. (eds.) *Handbook of Disability Studies*. London: Sage, pp. 123–144.
Wittgenstein, L. (1970): *Über Gewißheit*. Frankfurt am Main: Suhrkamp.
Wolf, S. (1994): Happiness and meaning. Two aspects of the good life. In: *Social Philosophy and Policy*, 14, pp. 207–225.
Wolf, U. (1999): *Die Philosophie und die Frage nach dem Guten Leben*. Reinbek: Rowohlt.
World Health Organization and The World Bank (2011): *World Report on Disability*. Geneva: WHO.

# Part I

# Norms and Body

# 2
# On Unfamiliar Moral Territory: About Variant Embodiment, Enhancement and Normativity

*Interview with Jackie Leach Scully*

Christoph Rehmann-Sutter: *One of the recurrent topics in your work is the philosophical and ethical trouble that arises from the fact that we humans have bodies that vary to quite astonishing degrees, and at the same time we live within a culture that neglects this variability and instead cherishes an ideal of a human body. The body is mainstreamed by certain norms that we can talk about and that is important to reflect on because, as I understand you, enhancement as a biotechnological idea and strategy could not be understood without knowing about these normalization mechanisms. But before we go into the details of these mechanisms and their implications for ethics, I'd like to ask you why you see this as a problem at all. Is it not something that happens in every known society: that it has its ideals of a good or even better human body, one way or the other? People want these ideals as something to strive for in the way they work at their appearance. Look at beauty contests, the fashion business.*

Jackie Leach Scully: Of course it's true that all cultures seem to have an idea of what kinds of body there ought to be, if I can put it like that. But I would be a lot more cautious about saying that all cultures, past and present, have the same *sort* of idea about it. I think we can very easily acknowledge that people in other societies might have, or have had, notions about physical beauty or desirability that are quite alien. What may be harder to grasp is that in other societies, the very concept of a physical ideal might have significantly different parameters: for instance it might come with weightings of obligation, pride, desire and so on that are quite unfamiliar to us, in our culture and our time.

CRS: *Can you give an example?*

JLS: I'm thinking here, for instance, of the way in which pre-modern European cultures seem to have been intensely oriented towards a sense that 'rightness' in the way a person lived was demonstrated through their place in the social structure, and also their place within a religious or spiritual structure, rather than through achievement of a physical ideal. I'm not of course suggesting that the medieval world didn't have physical ideals, but that the investment in them may have been more diffuse. Or to give a more contentious example: the extremely racialized bodily ideal of National Socialism; here, the body had an overtly political significance unlike anything we are used to today.

Clearly people do seem to accept (I'd hesitate to say 'want') that these bodily ideals are present and powerfully active in society. And of course, up to a point having something like this to aspire to is a positive thing: you'd have to be a real curmudgeon to argue that taking pleasure in one's appearance, at least to an extent, is something to be disapproved of! The important thing to hold onto, though, is that sense of 'to an extent': I'm suggesting there is a point beyond which an interest in one's appearance becomes disproportionate. It turns into vanity, which is an old fashioned and little used term, but one that I think is salient here.

So one problem, probably the one I'm most concerned about, is not that people have these interests and aspirations, but that they get out of hand. This is exacerbated when practical technologies become available for modifying/enhancing bodies, and those technologies are then directly and indirectly driven by powerful commercial forces. And on the whole, commercial forces are highly conservative: for obvious reasons, they have strong interests in ensuring that people end up wanting what they, the producers of body-modifying technologies of all kinds, can provide.

CRS: *I know the topic of beauty remains a bit superficial philosophically, when we start from the fashion business, but it is still an evident social phenomenon that plays a big role in the lives of many people. What's the deeper meaning of beauty in your view?*

JLS: I wouldn't want to describe 'beauty' in itself as superficial – after all, it's long been a legitimate topic for philosophers and artists, as well as fashion designers and engineers and scientists! And as you say, wanting to look beautiful or fashionable is a social phenomenon we can't ignore. It's also very complex: as I've mentioned already, there are social scientific questions to be asked about what constitutes beauty, how people know about that, how much deviation from the ideal will be tolerated, what are the normative limits of aspiring to our ideal, and so on. For me, the ethical issues are strongly tied up with the power of these social forces. I don't think that someone paying £10,000 to have their eyebags cut away is doing a grave moral wrong (though I would say they are doing *some* moral wrong, given that an investment of £10,000 could benefit a lot of people, for much longer than a facelift lasts, in poorer parts of the world). I have more serious questions about the moral culpability of the medical professionals, the media, advertising, and so on, who put their lives to the service of convincing a person that the path to happiness is to have his or her eyes done.

CRS: *When you say the beauty of the body is a social phenomenon, what does that imply? There are norms involved, quite obviously. But there is also something else. Beauty is dependent on the kind of relationship. Somebody can find me beautiful, regardless of the norm I am fulfilling or not fulfilling. Other people would certainly find me unattractive. And this also happens to people with so-called disabilities. What I want to suggest is that beauty, attractiveness, as well as feelings of shame or pride, must somehow have to do with the sociality of our bodies – or the embodiment of our relationships. Is there an ethics within this? What is going on when somebody finds somebody else beautiful?*

JLS: What we find beautiful, or even what we find normal, ugly, repellent, and so on, is a social but also a social-psychological and psychological phenomenon – and in many contexts a political one. To come anywhere near answering the question 'what is going on when we find another body beautiful?' would involve exploring all those domains, I think. Clearly there is something about fulfilling or not fulfilling norms that is, in a sense, superficial, because it can become irrelevant when set in the context of a particular relationship ('she may not have been conventionally attractive, but she was beautiful to me' – that kind of thing). I think you are right that what we experience aesthetically, if you like, is an embodied relationality. Where there is virtually no relationship to speak of, then perception is probably more 'objective' in certain senses. There are some interesting points to explore here about the way in which our perception of deviations from a physical norm can be heightened or minimized, depending on our relationship to the person involved – that, psychoanalytically, our relationship is with an internal representation that reflects the 'reality' more or less well. I think that can actually take us somewhere quite ethically uncomfortable, if what it means is that in order to find a physically anomalous person acceptable, we unconsciously stop seeing the full extent of their unusual morphology. But that may be something that the human psychology of perception just does. The social aspects of physical norms and ideals (that is who gets to set them, how rigidly they are policed, and what sort of political and commercial interests underpin them) are more troubling because they are, in principle at least, contingent and open to change if they are ethically problematic. In which case, we have to acknowledge a moral responsibility to change them.

CRS: *In your book* Disability Bioethics *you write that we can 'think through the variant body'. Can you explain the basic idea in this?*

JLS: In that book I was using the idea to cover several slightly different ways in which we can usefully reposition the body in processes of ethical, and other, thinking. Perhaps the easiest or most trivial one is simply to take the body more seriously as the arena where moral good and moral harm are played out. It's become almost a reflex today to criticize moral philosophy for being somehow disembodied – that in trying to make universal claims, its tendency is always to move away from the specifics of body and social embeddedness as rapidly as possible. This has become an almost boring point to make, and I think those of us who argue it tend to forget just how much ground there is still to be made up. To put this more concretely, medical and clinical ethics have always tended to prioritize instances of major decision-making as being 'real' ethical issues and have paid less attention to the ethics of everyday interactions between particular bodies in a healthcare setting. The latter are a good deal less dramatic – courtesies of touch and recognition, for example, not mercy killing or abortion! But they are also harder, in the sense that they aren't easily placed within existing ethical frameworks.

Another meaning of 'thinking through the body', which I elaborated in more detail, is to do with what the body means for how people think. Again, it's a bit of a caricature, but there is a sense in which the model of the brain as the location of thinking so dominates our picture of ourselves that the bit below the skull becomes almost superfluous: a handy way of shifting the brain from place to place, and funneling information about the world to the neurons, but not good for much else. Part of that is true: you don't do a lot of higher order cognition with your little finger alone. But equally, it's important not to lose sight of the way in which the higher processes of the brain ('thinking') are embedded within physiological systems that don't just *influence* a function like perception: in a very real sense they are what *make perception happen*. And those physiological systems are distributed around the body (and also beyond it, once we take into account the effects that implements and technologies have on how the body interacts with the non-self world).

I started to puzzle about this really because of the observation that, sometimes, people with the kind of variant body that we called disabled could hold ethical opinions that were different from those held by standard model, or non-disabled, people. I want to be careful here to say that it wasn't that these opinions were radically distinctive or ubiquitous, but that in certain cases there were quite marked differences in priorities, values, and judgments. I felt it was too easy to respond dismissively with 'oh, they would say that in their position', or slightly more positively with 'their experiences have led them to think that'. I wanted to join the dots a bit more fully. What, exactly, is it about experience that leads people to think differently? In what way does the experience, as it were, get inside someone's head? (And there of course I am doing precisely what I've just criticized philosophy for doing, that is arguing as if all the thinking goes on between a person's ears!) Clearly this is a general point about epistemology and ethics. I was interested in disentangling the different ways in which a disabled person's distinctive embodied experience – distinctive because she has a variant and not a standard body – could be part of her ethical perception and judgment.

In the end I drew on a cross-disciplinary set of ideas, ranging from the phenomenology of Merleau-Ponty and others, through work in cognitive neuroscience on embodied and distributed cognition, to the anthropologist Pierre Bourdieu's writing about habitus. All of these are struggling with approximately the same question: how does who you are, physically and socially, affect how you think? It's an extraordinarily difficult area because we lack the analytical methodologies and techniques needed to join those dots; we can only come at the question indirectly.

CRS: *I understand that this set of ideas originates from your work about disability. When we now consider biomedical interventions, such as genetic ones, which are usually classified as 'enhancements' (in the sense that they go beyond restoring some dysfunction or loss of function), what are the implications of thinking through the body?*

JLS: The implication would be that such interventions, by changing the nature of the body and therefore modifying its interaction with the world, could change the way in which that person thought. I'm not being specific here about how thinking might change, or how much; as I've just indicated, there is very little useful data to support claims of that order. And if any differences are subtle, how would they be detected? This would be a particular feature for genetic enhancements which, if they were germ-line modifications, or were introduced *in utero* or very early in life, would then be that individual's norm. The effects of interventions later in the individual's life, or of non-genetic enhancements – pharmacological, sensory, involving a prosthesis, and so on – might be easier to track. But still there would remain the issue of knowing what sort of effect to look for, and where.

I realize this description makes ethical perception and judgment sound more mechanistic than I would want to. And I'd also want to say that I'm not assuming that altered thinking through a variant body as a result of biomedical intervention is necessarily something to be deplored or feared. I *am* saying it's an area worth exploring, particularly because it could mean that enhancements come with a cost (physical, cognitive, emotional, social) to the individual. Much of the ethics of enhancement so far has worked from the premise that, as long as the technology doesn't inadvertently harm the individual, then by definition an enhancement must benefit *her* even if there are ethical costs to society, for example through the distortion of healthcare provision, through increased expectations or by exacerbating social divisions. But if enhanced bodies have more diverse effects on the individual than straightforwardly improving a function, then there are also likely to be ethical costs to the individual – and that would change the balance when weighing up harms and benefits of the intervention.

CRS: *This is a very important concern, which is rarely seen in ethical discussions of enhancement technologies. We normally just take for granted that people with improved bodies would make their ethical judgments in the same way that we, the non-improved humans, make them when we evaluate what will be an improvement and what would be harmful or risky. If I understand you correctly, you hold that the improvement of body functions, such as amplification of cognitive power, or postponing the process of aging, could possibly affect the moral world of these individuals. This follows from your reading of Merleau-Ponty and other phenomenologists. And because we cannot foresee in what way an enhanced person's ethical perception and judgment will change, we, the non-affected, therefore cannot make a reliable moral assessment of the value of such a change for them. Hence, improvement must be conceptualized in a radically perspectival way. It matters for whom something is supposed to be an improvement, because their bodies may differ. And in regard to plans to change embodiment 'for the better' we cannot foresee how those affected will in fact evaluate the changes in their bodies, and whether they will indeed be betterments for them. This makes an ethical assessment of enhancement technologies quite difficult, if not impossible. And those who 'live' the change don't know how it would be to live as another embodiment. For them it is just the norm, as you say. They will adapt to their embodiment.*

*There are some difficult questions in that. The first one is obvious, how should we, as non-altered people, judge that the different moral perception and judgment of a person with altered embodiment is in any way worse or deficient? It will be just different from ours. What right do we have to say that their adapted moral perception, or their feeling of being 'right' as they are, is problematic?*

JLS: That is simultaneously a very trivial and a very profound question! You're right in the sense that if the moral perception or judgment of a person with a variant embodiment is in some way different from ours, there is no *a priori* reason to evaluate it as a worse or less adequate judgment than ours. It is, as you say, just different. But in reality, there are many situations in which we don't simply stand back from those judgments and say, well OK, your viewpoint is just different. We make evaluations of different perspectives: Are they reasonable? Are they coherent with other stances and values? Are they something we can tolerate under the broad umbrella of whatever a modern, democratic, and pluralist society can live with? Or do they just go too far? And why are these areas of judgments, specifically, and not others, under scrutiny as to their acceptability?

From that point of view, such questions feed into a very large debate about the parameters of shared moral understandings in societies that are increasingly socially, ethnically, and culturally diverse, and where alternative moral understandings to the mainstream ones are becoming more prominent – one might say, gaining in political and moral confidence. So on a theoretical level these debates are far from new. What *is* new is the possibility that, as well as different moral understandings arising out of distinctive cultural, religious, or ethnic milieus – the sort of groupings that have long been considered to constitute discrete individual and political identities – they may also be generated by more contested sites of difference to do with the body and embodied experience. The question then becomes, how do we identify and evaluate these judgments?

In this context it is important to understand, as far as is possible, what particular acts and practices mean for different groups, and not to assume that meanings can be extrapolated in a facile way from one group to another. To return to the example that I have been looking at, on and off, for some years: when signing Deaf people indicate that they might, conceivably, wish to use assisted reproductive technologies to 'select for' deaf children (or at least not to select against them), we need to have a view of the full context within which such an evaluation might be made – the context of a flourishing signing Deaf community, for example – in order to tell whether that decision is in itself consistent with the prioritization of the child's welfare or not. Deaf and hearing people will not disagree with each other about whether a child's welfare should take priority. But where they may differ (and this is what is affected most strongly by their differently embodied experiences) is how 'the good of the child' may be acted out. As a result, the Deaf community can see a preference for a deaf child, or at least the absence of a preference for a hearing child, as fully in line with the desire to protect the best interests of the child, the parents, and the community as a whole; where the hearing world are likely to perceive this choice as actually harming the child's best interests.

Ultimately this means I think that acquiring baseline empirical knowledge becomes an ethical requirement. To understand whether a 'different' moral evaluation is 'different but understandable' or whether it means 'this is beyond the limits of what is comprehensible within the common moral framework', we need to have a clearer knowledge of what various bioethical decisions actually mean to different actors in different contexts.

CRS: *Even if we agree with your argument about variant embodiment in principle, aren't there nonetheless some differences to take into account? Obviously not all body changes will be equivalent. Can we differentiate between ethically neutral and possible non-neutral body changes with regard to changes in moral perception? One that could fall under the 'neutral' category and will probably – but yes, that is still a question – not change the ethical thinking of the person concerned, could be the improvement of resistance to diseases. If we could add some extra genes to our children that make them considerably less susceptible to cancer throughout their lives, much less than today's average cancer susceptibility, this would count as an enhancement. Their bodies would be stronger, more resilient. But the changes would not alter everyday life. But if somebody were to get a neuro-implant that enabled her to instantly memorize everything that she reads, it would affect everyday life, and very likely also ethical judgment. This would be an example of a non-neutral body change that may affect moral perception.*

JLS: That's an interesting question. Of course, not all body changes are equivalent in any sense, and certainly not in the sense of affecting moral perception. I'm interested by the example of the 'neutral' category that you give, though, because in fact one can imagine this actually having profound non-neutral consequences for moral understanding. For instance, improved resistance to disease might have the consequence of making people feel less empathy with those who do still show vulnerability to disease. It's quite imaginable that in such a case, the enhanced person would not feel that she or he has experienced astounding privilege, but rather that people still subject to cancer or other kinds of disease are just weak, repellent, to be shunned. (It would be nice to imagine that kind of constitutional superiority going hand in hand with increased compassion and sense of gratitude for the luck one has experienced, but unfortunately social-historical precedents don't give much cause for optimism here.) So in a sense, this kind of enhancement would actually have very profound impacts on areas of everyday life that don't have anything to do, directly, with cancer.

I agree that the neuro-implant case would have more immediate and direct consequences for everyday interactions, with animate and inanimate others, that inform subjectivity. I'm not sure, though, that the situation you describe would necessarily give rise to much more radical changes in moral perception or judgment than having significantly enhanced disease resistance.

One of the messages of *Disability Bioethics* was about how bad we are, collectively, at predicting what the significant consequences of a new technology are going to be. In that book I was thinking specifically about predicting what were the likely consequences of a disabled embodiment on moral understanding, and also the possible consequences of a biomedical intervention or assistive technology on the life of a disabled person. However, the idea can be applied more generally than that; and the conclusion, more generally, must be that there is an absolute necessity for empirical and experiential research into the real effects of such modifications. We can make all the predictions in advance that we like, but a growing track record of experience with a range of new social technologies suggests that the real changes, problems, and challenges are likely to be both more mundane than expected, and unpredictable. In the face of that kind of dilemma we need to swallow a large dose of epistemic humility, and explore the empirical reality of embodied moral understanding.

CRS: *Let me ask you one key philosophical question, just to clarify how we think, do you suggest that we should believe in a form of embodied relativism? Or are there still reasons to believe that some basic ethical concerns, for instance the ideas of injustice, discrimination, or exploitation, must be universal, regardless of the kind of bodies we have? In an enhancement world we could end up in a situation like Deckard in Ridley Scott's movie* Blade Runner, *who falls in love with a replicant but is supposed to hunt her down and kill her. The language of the emotions clearly tells him that this would be a bloody form of discrimination, regardless of her ontological status as artificial. You just need to replace 'replicants' with 'improveds', and you can pose the same question. Don't we have strong reasons to defend a universalist approach when it comes to oppression?*

JLS: I'm quite sure I'm neither advocating nor diagnosing the kind of relativism you are indicating here, embodied or otherwise. First, there are some very basic ethical concerns and concepts that are clearly shared within many societies, and may well be universal: these are things of the order you mention, such as a concept of (and rejection of) injustice, the idea of discrimination, giving special weight to the welfare of children, protecting the vulnerable, and so on. The important point here is that although the outlines of these concepts are shared, ideas about how they are operationalized – that is, what acts and choices constitute discrimination, against whom, and so on – may not be. That level of idea works much more, I think, on the level of moral intuitions that are acquired through socialization processes of various kinds, within the family, school, and wider community.

In *Disability Bioethics* I was suggesting that a person's embodiment might contribute in a variety of ways to the formation of their moral intuitions. I considered that this might happen in terms of the kinds of experiences they had: that some might be distinctive, or have distinctive features, as a result of having a particular embodiment. An obvious example here would be pregnancy as a function of gendered embodiment, but there would be others that derive from having a disabled embodiment – having a prosthesis fitted, for example. These sorts of differences are fairly easy to see.

Less obvious perhaps is the possible contribution of embodiment to the meaning people make of experiences, even ones that are shared. For example, that the meaning of testing for anomalies during pregnancy might be different for a disabled than for a non-disabled woman. I don't suggest here that the moral meanings of this experience would necessarily be absolutely different for a disabled woman, and certainly not that a disabled woman's moral understandings in general would be radically unlike those of a non-disabled woman. It was simply the claim that such embodied experiences generate a sense of what is obviously right or obviously wrong, the intuitive responses that people will then go on and produce better or worse justifications for, if pushed.

So I was making what I think is the fairly uncontentious claim that, within a framework of basic ethical values and beliefs that is generalizable across a society, and may even be universalizable across all societies, there are differences in how those values and beliefs are considered to be lived out; and that it's within this kind of moral register that disabled or enhanced embodiment may make a difference. It's less of a normative argument about relativism or universalism than a hypothesis about where differences in moral intuitions might come from – and especially a pointer to the neglect of the body as a possible source of difference.

CRS: *Yes, I can better see your point, and I enthusiastically agree about the importance of the body. But let me press you a bit on this. I still ask myself, how the different embodiments can matter, how they enter the moral worlds. How can we distinguish between values, which may be roughly the same across all societies, even universal, and the differences in how those values and beliefs are considered to be lived out, or the moral intuitions that different people with different bodies develop? Both are kinds of values, both are socially and culturally formed. And both are necessary in order to fight oppression and injustice. I don't see how universal values can work without the feelings and intuitions that support them.*

JLS: I think your second point is the key one here. I would agree that widespread or even global moral values, and more parochial ones, are socially and culturally formed; but clearly the register, if you can put it that way, is different. As I said before I think there are very fundamental beliefs about moral behavior that are culturally widespread: I can't think of a culture that endorses indiscriminate killing for example, or that doesn't hold the welfare of the child to be a good thing. Those are socially formed in the sense that the ideas of 'indiscriminate' and 'killing' and 'child' are social ideas. (It's true that the concepts of killing and of the child both entail a material or biological reality, but distinguishing between killing and death, and defining when a child turns into an adult, are social moves.) I don't want here to enter into a debate about where those very fundamental ideas come from, to what extent they are hardwired, or are behaviors that have been selected for because they promote the survival or flourishing of individuals within groups, or because they have some metaphysical origin. Whatever the origin of these values, they are of course held in place by the associated behaviors and intuitions. But I think they can be distinguished, albeit imperfectly, from the working-out of particular 'good ways of being' that embody those fundamental values in actual lives. Intuitions derived from fundamental shared moral values are what make certain practices, and not others, carry a degree of plausibility as a way for members of a society to live. It's complicated of course, because over time as practices turn into habits and traditions, they too become part of the intuitive texture of moral life.

How different embodiments then matter is the question I tried to explore in *Disability Bioethics*. One of the conclusions I came to was that because there are multiple ways in which embodiment might, in principle, modulate moral evaluations, we need a similarly diverse investigative approach – examining how bodies affect social interactions (we know a bit about that), the psychological processes through which social interactions might shape consciously held moral opinions (we know rather less here) and also unconsciously held ones (almost nothing known), and how variant motor, sensory, and perceptual possibilities – including those that involve assistive devices and extensions – can alter moral cognition). All of which is to say that, at the moment, people like me can devise all sorts of fancy theories about how embodiment might affect moral understanding, but don't have much in the way of empirical data with which to test out theory!

CRS: *Let us now turn to a different topic: the side effects of regulation. I am thinking of pressures that can be exerted on individuals and societies by moral or legal norms to regulate enhancement. In a paper of 2001 on human genome alteration, or 'gene therapy', which we co-authored, we claimed that it is unwise to regulate this field by a rule that allows therapies and bans enhancements, if the latter category is defined on the basis of a distinction between the species-typical or normal human functioning and other functional states that are beyond, above the normal. It is unwise, we said, because it will have discriminatory side effects for those who actually live with variant bodies. Is that still today a valid criticism in your view? And could you explain how you think that argument should be made?*

JLS: Things have clearly moved on since then – not least because developments in the life sciences, like work on neuropharmacology, mean that the drawing of a bright line between therapy and enhancement is considerably less plausible. Back in 2001, most of the debates around the ethics of enhancement started from the premise that we could distinguish clearly and straightforwardly between two classes of intervention. When we wrote that paper, what we were particularly concerned about was that enhancing and therapeutic interventions were being defined specifically in terms of their relationship to normality – so that therapy was restoration to the normal range of human form and functioning, while enhancement was going beyond that – and as a result we thought that if these interventions were then going to be regulated, something like a definition of human normality would begin to enter into law. And this would be virtually unprecedented, at least in terms of the scope and level of detail necessary to craft such a policy.

We also argued that it would entrench a very narrow view of human normality, and that this could be damaging to people with variant embodiment – that is, with body forms or functions that fall outside the species-typical range, some of which are identified as impairments. The harm would result from the way in which such an entrenchment ignored the possibility that an intervention which would be considered therapeutic by non-disabled people – restoring them to a norm – might be experienced by disabled people as enhancing – as taking them beyond a state of form or functioning that is their norm. This perspective would have to be ignored, in order for the intervention to be considered a therapy and therefore be legal; but doing so would effectively be saying that disabled people's capacities for self-definition and self-determination are not to be taken seriously. (Doing so would be depressingly in keeping with a history in which disabled people's rights to self-determination have been systematically ignored, but historical precedent is hardly a point in its favor.)

Today we have moved to a general recognition that many biomedical or biotechnological interventions don't fall cleanly within the categories of enhancement or therapy. There are too many cases in which interventions could both 'take the individual beyond normal' and 'restore to some kind of normal'; for instance, a pharmaceutical developed to ameliorate the memory loss of dementia could also, at different dosages, enhance the capacities of someone with standard memory. If there were no differences between the 'therapeutic' and 'enhancing' uses in terms of harmful side effects, it would be difficult to argue that one use is permissible but the other isn't, without also arguing more generally that the statistically normal range of body forms and capacities should be normative. There are ways of arguing for this, of course, but I think it's clear that neither they, nor the transhumanists' arguments in favor of a much more liberal line on the transformability of the human body, are universally compelling at the moment. If they were we wouldn't still be debating them so much.

One thing it is important to keep in mind (and is sometimes hard to remember for those of us who spend so much time discussing these issues) is that we are, socially and culturally, on unfamiliar moral territory here. Until very recently, individuals or societies have had few means available to control the kinds of bodies that they or others have. Societies have always had opinions about how much physical or behavioral variation is tolerable, and what constitutes an impairment, but actually enforcing those opinions has been restricted to crude methods, like infanticide or the social ostracism of disabled people, which over time have become increasingly unacceptable. What we have now is a situation of growing technical ability to select for the phenotypes we want by prenatal, preimplantation, or pre-conception interventions, or to change the capacities of existing bodies by pharmacological, neuroprosthetic, or other means. And this, I'd suggest, puts the societies that have access to those technologies in the potentially dangerous position of being able to act on historical assumptions and prejudices about bodily variation that have never really been subjected to proper scrutiny.

It's a fascinating historical moment. These selective and manipulative technologies are coming on stream in a social and political context which in some ways is more accepting than ever before of diversity in general, and of disabled people, and where disabled people are beginning to have a global political voice as well. Yet at the same time, as we discussed earlier, the cultural context is one in which the pressure to conform to particular standards of physical appearance and behavior is mediated through extraordinarily effective commercial, governmental, and bureaucratic agencies. Future

> forms of bodies, and the kinds of lives that can be lived by them, will be determined by the bioethical deliberations going on today, because they will lead to policy decisions and to the creation of a cultural atmosphere that is more or less hospitable to diversity; and that's why it's so important that we engage in these deliberations with care.

The interview was conducted via email in 2012. Jackie Leach Scully refers to her book *Disability Bioethics. Moral Bodies, Moral Difference*, which appeared in 2008 (Lanham: Rowman & Littlefield). The article by Scully and Rehmann-Sutter mentioned is 'When Norms Normalize. The Case of Genetic Enhancement' (*Human Gene Therapy* 12, 2001, pp. 87–96). In 2002, Scully's Swarthmore Lecture appeared as a book entitled *Playing in the Presence: Genetics, Ethics and Spirituality* (London: Quaker Books).

# 3
# Improving Deficiencies? Historical, Anthropological, and Ethical Aspects of the Human Condition

*Christina Schües*

Reproductive medicine, biotechnology, and neurosciences provide the technological means for the enhancement of bodily and mental capacities. Enhancement is an intervention into the body and the self that concerns and alters a person's self-understanding and self-actualization, and thereby the *conditio humana*.[1] The human condition consists of particular conditions and features such as age, natality, mortality, gender, worldliness, vulnerability, and the need for nutrition and support; and also, generally speaking, 'disability is part of the human condition', if not permanently then at least temporarily.[2] We can experience these features, but they are not necessarily directly visible, like being mortal or vulnerable, or having a predisposition for a particular illness. How or when these features can be experienced also depends upon – as I call it – the *conditio mundana*.

The *conditio mundana* is the constitution of the order and materiality of the world in which we live. The constitution of the *conditio mundana* influences human life fundamentally and consists of worldly features, such as the factual or normative ordering of human relations, the economic or technological systems implemented, the architecture of objects or the ecological infrastructure of the environment, as well as cultural and social implementations of time and space, norms and values.

Modern medical technology is not only directed towards the therapeutic benefit of sick individuals; reproductive medicine, gene technology, and neuroscience in particular are also concerned with the improvement of bodily and mental functions and features of single individuals or even of humankind. It is not always clear whether the improvement of an 'ability' is supposed to be a shift from 'disability' to 'ability', or from 'ability' to 'better ability'; it also remains questionable what 'improvement' actually means. Medical and technological interventions take place in particular material and social surroundings and they may have consequences for particular human conditions as well as for the order and material basis of the world.

The notions of therapy, enhancement, and the general improvement of human beings and their bodies cannot be clearly separated. Below, I take 'improvement' as the guiding notion because it seems the broadest, encompassing therapy as well as enhancement. Both of the latter terms have to do with improvement, and both depend upon the question of how such improvement is placed within the life of humans, without turning to strict criteria for health, normality, impairment, or other references; but in general, therapy, improvement, and enhancement are 'overlapping categories' of human interventions with the goal of enhancing human capacities or characteristics.[3] The term 'enhancement' is mostly used in the following areas: anti-aging and, for instance, possible intergenerational conflicts because of longevity; genetic interference that, for example, affects future generations; bodily and neuro-enhancement of, for instance, cognition or emotion; and the human/machine interface created by implanting biochips into the brain.[4]

Most researchers phrase questions about the legitimacy of enhancement in three ways. First, should the task of medicine not be limited to healing or preserving the abilities or characteristics of humans (and future generations)? Are we instead 'playing God'? Second, is it economically reasonable to use the resources in the health sector for enhancement practices? And third, is it ethically legitimate in respect to justice, resource allocation, or equality of chances?[5] These questions, however, limit the discussion as if only one set of characteristics of an individual were enhanced, and thereby transformed, without also affecting that individual's whole self-understanding, as well as the normative settings, and even the *conditio mundana*, by which I mean the concrete material and social conditions and the world order in which people actually live. If humans and their conditions are transformed substantially, that is, if humans are enhanced, then the order and materiality of the world will not remain the same. The surrounding world, the life-world, society, and culture are not stable containers; they change and are transformed according to scientific understanding and technological implementations. Therefore, questions must be posed about the constitution of the world in respect to the *conditio mundana*.

Researchers of enhancement technologies generally exclude the question of adaptation: what gets adapted to what? Is a technology used to improve the human condition *for* a given technologically designed world?[6] Or, perhaps, should we 'improve' the world's relations and structures, to provide better support to humans as they are? The focus on the thematization of the human condition and on questions of legitimizing enhancement implies: in what sort of world do we want to live? How should the world be designed in order to provide the conditions for a 'good life' while respecting the individual's specific human conditions? These are profoundly ethical questions. In order to clarify them, the motives, meanings, and consequences of enhancement and improvement need to be elucidated both historically and systematically. The motive for improvement has a long tradition in relation to the deficient human; is enhancement simply a derivative of this tradition? What were the motives for improvement? The first part of this essay reflects upon these issues. This historical attention is needed in order to define, in the second part, some characteristics of 'enhancement' and address the *conditio humana* with reference to its different features and conditions, questions of necessity and contingency, and transformation. In the third part I will explore the correspondence between the transformation of the human condition and the shaping of the world – the *conditio mundana*.

## The history of the deficient human and different motives for improvement

In occidental philosophy, humans have traditionally seen themselves as deficient, and hence have been dissatisfied with themselves. Deficiency has been noted since Greek antiquity;[7] however, the consequences of this observation have been rather different. For instance, Plato's Protagoras describes in a myth the human being as 'naked, unshoed, unbedded, unarmed' (Plato (1967), Protagoras c. 321c) and living in a hostile environment; by contrast, Epicurus (2011) also recognized such bodily deficiency but considered it to have no deeper meaning. For him the meaningful challenge for humans is of overcoming the deficiency of fear or too strong desires. In different times, generally speaking, human self-dissatisfaction has been grounded in physical limitation, bodily or mental capacities, human susceptibility to diseases and illnesses, the inevitability of aging and death, and the lack of moral character. The improvement of humans was supported – at least in European history – by clothing, prostheses, gymnastics, or scientific or ethical education, from Socrates via the century of the Enlightenment to the present day; by race-theoretical and biological fantasies of breeding in the totalitarian states of the 20th century; or by physical practices such as prostheses developed through the technologies of physics, medicine, and mechanics from the 17th century on. Improvement has also been sought by means of cosmetics, cosmetic surgery, and different kinds of nutritional elements or drugs.

The notion of enhancement is often derived from the concept of improvement. The report of the President's Council on Bioethics calls enhancement the means of 'improving native powers'.[8] But is it convincing to infer the concept of enhancement from improvement? And is it true that humans always have a motive for improvement or even enhancement? Is better always good? It seems to me that the following examples of historical motives for improvement have been forgotten, because they seem to be taken for granted today, and hence reflection on them is neglected.

## Ancient philosophy – *Eukrasia* and harmony: Virtuous men and the good life

Socrates saying that he knows he 'knows nothing'[9] marks the beginning of the history of the human conceptualized as mentally deficient from birth.[10] Humans are certainly more deficient than gods or angels. The idea here is that if humans realize that they lack something, they try to compensate for this lack; for example, when you are thirsty, you drink something; similarly, when you realize your lack of knowledge, you try to learn. Thus, *lack* is understood as deficiency and motivates improvement. In the Apologie, Socrates forcefully calls upon his fellow citizens to care about truth and wisdom and thereby concern themselves with the 'perfection of the soul'.[11] Socrates' acknowledgement of the deficit – 'I know nothing' – already posits the human in-between someone who believes that she or he knows something, and hence feels no urge to strive for any improvement, and a god or angels who are perfect by definition, and hence do not need any improvement. Thus, someone who acknowledges a deficit also realizes the need for improvement and is already raised above 'normal' people who are ignorant because they think that they are already complete. Deficiency becomes the motive to improve oneself.

Improvement in this context means acquiring better knowledge, and thus participating in truth. However, becoming knowledgeable does not mean simply being quantitatively filled with more information; rather, it means a qualitative move from a disordered sensible world to the world of ideal concepts, to the original of all objects in the daily world. And only here, in the orderly world, which is devoid of chaos, illusion, or misunderstanding, can humans reach truth and happiness, and hence a good life.

Plato's concern is directed towards the virtue of humans because he believes that any conception of the *polis* (state) corresponds to the virtues of human beings. In ancient thinking the healthy organism,[12] the *polis*, or the soul in harmony resembles a well-ordered cosmos.[13] However, the world order itself is static and only considered within the limits of ordered or disordered human action and cognition. A physician is the doctor for the body and should care for its harmony; a philosopher is the doctor of the soul, and hence it is his task to free the soul from disruptions or confusion. If the order or the figuration of the elements is disrupted, it is the task of medicine to refine the balance. Refining the balance of the elements means changing the deficiency (*dyskrasia*) to harmony (*eukrasia*) in order to (re-)install the health of the organism.

*Eukrasia* is a quantitative concept, used by medicine, but it is understood in terms of the quality of a well-ordered body.[14] In the dialog Philebus, Plato (1925a) uses a similar way of explaining the harmony of the soul. He discusses whether or not *aisthesis*, that is, all the senses, should be taken inside the concept of *eukrasia*. *Eukrasia* has the prefix *eu* (good, well), which implies the idea of the good life, a certain harmony that cannot simply be explained by reference to normality.[15] A good life need not be a normal life! A good life, the *telos* of all action, is – for Plato as well as Aristotle – the life of the philosopher; or, speaking of ordinary people (who will not become philosophers), the highest aim may be to live a just life (*eudaimonia*), a life that would be worth being lived again.[16]

Any kind of improvement, let it be of life or a thing, can only be made according to nature and by emulating nature. Hence, even craftsmanship – *techne* – cannot achieve more than what is already inherent in the structures of nature. 'Nature guided the *techne* of medicine, men would simply imitate nature's principle, and any act beyond it was not imaginable.'[17] Later, in the Middle Ages, the Latin description of the *restitutio ad integrum* – the restoration to the original condition – is appropriate to point to the basic goal of human action: an order according to God (*Gottesverähnlichung*).[18]

## Modernity – Function and optimization: *Homo faber* and the calculation of happiness

The focus on consciousness, self-consciousness, and physical laws that affirm generality and apparent context independence are central to modernity. The understanding of human beings as machines, such as the provocative representation by La Mettrie (1999) in *L'Homme Machine*, opened the way to a technological interpretation and to optimizing nature beyond Platonic imitation and the Aristotelian mimesis principle. The strict dualism between *res cogitans* and *res extensa* in Cartesian thinking and the scientific focus on mechanisms led to the radical reduction of bodily material to issues of quantification and control. From then on physical laws, understood as 'analytical geometry'[19] of kinesthesis, controlled the extensions of material bodies just like functional machines.

*Homo faber*, the human who is guided by mechanical laws, is the stepping stone to the functional rules of cybernetics; and this means that not only the physical body but the whole lived body, that is, the physiological and psychological life of the body, such as emotions, desires, and moods, can be determined and calculated by physics and mathematical and computational modeling. As Hannah Arendt – having in mind political and ethical thinking – says harshly, the stupidity of the 'pain–pleasure calculus' has its roots in this thinking.[20] Thus, in this context, deficiency means dysfunction; therapy means the repair of machines; improvement is possible as the optimizing of a well-functioning to a better-functioning machine. Nature is no longer understood as a well-ordered cosmos but as a functioning machine, which can be substantially influenced if one knows enough. Hence, knowledge achieved by means of exact observations, for instance by dissecting corpses to understand physiology, or through scientific experiments, is the key to progress. In principal, nature has no limits and is open to be transformed to the optimum.

## Improvement of humanity and perfectibility of reason

In the 18th century most authors were concerned with humanity, understood as a virtuous attitude, based on the notion of dignity. Humanity was found in humanism, being human, human duties, and dignity, and is defined in strict contrast to its opposite, understood as the animalistic, the instincts or desires. The concept of humanity was also directed against all doctrines that reduce the human to nothing more than part of a natural or historical process. Hence, any improvement in humanity is supposed to be more than natural; humanity refers to a general determination of humanism that extends over all differences such as race, cultures, states, or human dignity.[21] Human deficiency means a fall into beastly nature.

Most importantly though, the notion of humanity was used to declare the *telos* of humankind: humankind has to develop its emotional and mental dispositions as if it were an organic development. If the human is considered as an end, then the meaning of life must necessarily be optimally realized and human entelechy must be fully unfolded to general reason. Entelechy means the teleological force of a living being to develop itself according to its natural disposition. The notion of perfectibility was used to describe the teleological sense of the human being: education of the heart (Jean-Jacques Rousseau) or of reason (Voltaire). Other philosophers of the Enlightenment, such as Immanuel Kant or Moses Mendelssohn, demanded the factual perfectibility of the human species as a being of reason; only improvement in the use of reason will bring out humanity.

Kant locates humans between responsible maturity (*Mündigkeit*) and humiliation (*Demütigung*). In his 'pragmatic anthropology' (1977a) Kant raised the human above nature by turning him into the slave of nature at the same time. Humans are masters because they are members of the realm of ends and reason, and are supposed to be responsible, free, and autonomous, that is, mature; and since humans have this status they are deeply humiliated by also being a member of the realm of natural laws. Hence, improving, development, means overcoming causality and controlling behavior by reason and action. By becoming a member of these two realms (reason vs. nature, causality) humans are raised very high and have fallen rather low.[22] Hence, it is not just his body that is part of nature; it is the human himself who belongs to the two realms and, hence, needs to be perfected.

The other representative of the Enlightenment, Johann Gottfried Herder, used the notions of deficiency (*Mangel*) and circle (*Kreis*[23]) in his work on humanity in order to direct his program against the Cartesian scientific ideal of context independence and generality. His anthropology relied on some ideas of ancient philosophy,[24] was very important to the anthropologist Arnold Gehlen, and was heavily criticized by Kant. Kant's pragmatic anthropology posits the human being as a freely acting individual; Herder, by contrast, develops a 'physiological anthropology' that conceptualizes the naturalizing of humans according to their senses and forces by referring to the body's nature. Thus, he recognizes an 'eternal process of the organic creation which is embedded in each living creature'.[25] This creation comprises the whole human being: from the physiology of stimulus up to the abstract acts of reason. Herder is interested in this context because he focuses on the whole formation of human beings as a progressive development, an entelechy, by unifying humanity and physiology, nature and culture; humanity is the human's character but it is only an innate disposition; therefore, it must be developed concretely and with sensitivity to the context. Thus, the concept of improvement encompasses the whole human being and his context. The development to and formation of humanity is 'a production, which must be endlessly progressed, otherwise we sink back to rough bestiality, back to brutality'.[26] The threat of deficiency is bestiality; the goal of improvement is to be freed from nature.

Herder also situates the topic of human deficiency in a comparison with animals; his interpretation still influences philosophers and scientists today. 'That the human stands behind the animal in terms of force and certainty of instinct' can be taken as certain; however, 'lacks and deficiencies cannot be the character of his species'. At the 'center of these deficiencies' must lay the 'germ for compensation'.[27] Humans do not, unlike animals, have circles, a 'tight and uniform sphere' in which they always remain;[28] they are freed from nature, they are free for the world (*Freigelassener der Schöpfung*) by means of 'reason',[29] but therefore they are also always undetermined. Friedrich Nietzsche (2010) picks this up later on: the human is considered to be the 'undetermined animal'.[30] Herder disassociates himself from any kind of theories of origin, but also from a general *telos* of humankind. The motive for improvement is to free the human from nature through nature. Hence, nature is both a force and a deficiency. The development of humanity is a guiding category, task, and force within a process of entelechy, and away from nature's deficiencies.

## Philosophical anthropology and ambiguities of the 20th century

The notion of a deficient being is mostly associated with the highly controversial anthropologist Arnold Gehlen. In the 1940s he defined – from his point of view, by simply following Herder – the human being as deficient because as a species he is physically and morphologically inferior to other species. Thus, the notion of deficiency is seen in terms of the survival of the species, and is used as a notion of comparison; Gehlen fictionally compares the human being with animals in order to show its special position. Human deficiencies and disadvantages are the biological non-adaptation to the natural environment. Deficiencies are of a physical kind, such as lacking strength, strong nails or teeth; and they are also of a mental kind, such as having only a reduced instinct. Therefore, any openness to the elements of the world is a strain for humans because they are easily irritated, and this is non-animal-like.[31] Thus, Gehlen argues that under 'natural conditions' human beings would have already been wiped out. In order to survive, humans fabricate – just as Prometheus did in the myth – a culture to act as 'second nature'.[32] The human must live with a prosthesis in order to improve nature, which only then can serve human life.[33] This reductionist notion of deficiency and its form of compensation theory was highly disputed at the time, even from within philosophical anthropology.[34]

The 20th century is ambiguous on the question of improving human nature. In addition to disputes within the discourse of philosophical anthropology, controversial discussions emerged concerning evolutionary theories (based on the work of Charles Darwin) and eugenic theory (Francis Galton), which focused on 'sciences of improving stock'.[35] However, being critical of teleological interpretations of evolution, as the anthropologist Helmuth Plessner was, does not mean being against improvement out of a recognition of deficiencies. Furthermore, today, comparing real or fictive bodies or achievements still motivates the desire for improvement. Advertisements for consumption work through comparisons: you will look better, be more successful, younger; in sum: improved! Consumers are approached by means of promises and the potential – offered by the media, internet ads, commercials – to compare their body, style, or self with some valuable product or technology that can be bought or used for professionalizing or socializing one's 'whole being'. As a result, individuals take for granted that they are formable and improvable; the body and even the self become the material that can be formed and improved, where improvement might mean reducing deficiency and achieving normality, or raising the normal or average to something better, from the point of view of the client, the physician, or other consulting persons.

The practice of individual improvement may lead to a special type of tiredness. Ehrenberg calls it the 'exhausted self': the self is depressed and feels the need for stimulating medicine (e.g., Prozac) in order to keep going beyond her limits, is ready for enhancement.[36] But sometimes improvement is not limited to the adult individual or some bodily improvement or reduction of deficiency. Since about the 1970s, improvement has been more radical: fundamentally transforming elements of the human condition itself.

Before I turn to a discussion of the relationship between the human condition and enhancement, let me summarize the results of the investigation into the history of the terms 'deficiency' and 'improvement' so far:

- Deficiency meant disorder and chaos (*dyskrasia*) in ancient thinking, distance from God in the Middle Ages, dysfunction in modern times, maladaptation in natural anthropology, and degeneration in evolutionary anthropology.
- The diagnosis of deficiency grounds the motive for improvement.
- The perspective of the 'deficient' or impaired human herself remains – at least for the most part – unattended and non-thematized.
- Improvement may be understood as leading to a qualitatively 'good life' or to quantitative optimization (calculation of happiness).
- The body is generally formable; it is seen as a mechanism like any other, and can be repaired and should be improved to its optimum.
- Humankind has a *telos*, an end: humanity means having the task of perfectibility (God, reason, creation, entelechy).
- Valuing or rating comparison (as distinct from a non-valued comparison: similarities and differences) is one basic (quantifying) means of diagnosing a deficiency, and hence of finding a hierarchy between individuals according to their measurable abilities.

These remarks about deficiency and improvement lead basically to three conclusions within the debate about enhancement. First, the improvement of humans has a long history; the idea of improvement is derived from the recognition of human deficiency. Second, the history of the deficient being is based on an ethical fallacy: deficiencies should be corrected as soon as they have been diagnosed, but if humans are in principle deficient beings, then the improvement becomes an ongoing imperative and promise – you can and must compensate for your deficiency to achieve individual 'happiness'. Third, the history of the deficient human being is characterized by unworldliness. It is not the world and the relations between humans that are the focus of improvement, but the human being's nature.

## Framing enhancement

The historical overview shows that many discourses are guided by the hidden presuppositions that enhancement is or should be improvement, and that as such improvement is to be desired. The notion of and the motives for 'enhancement' are based, perhaps, on these certainties, but particularly when it comes to more 'invasive' implementations,[37] the improvement is not always felt by those who are supposed to experience it.

Generally speaking, the notion of improvement implies bettering situation A by a certain means in comparison to situation B. First, however, the judgment that something is better implicitly demands objective criteria that it is indeed better for everybody. Second, we can argue that the very same improvements that improve things for one person do not improve them for another. Third, even though one might agree on the improvement, one might still disagree on the means of achieving it. Hence, by abandoning unifying concepts of humankind and its *telos* and by introducing individual approaches and contexts, and different points of view, the distinction between good and bad is sometimes useless; for example, in the case of procedures that concern implantations of technology into young children, those that affect future generations, conflicts in genetic diagnosis, transplantation practices, or neurotechnological experiments. The simple judgment of good or bad must be abandoned not because of a particular technology, but because of our ways of thinking and attitudes towards ethical decision-making. Improvement is thus not appropriate as the only conceptual basis for finding some definition of enhancement, for reasons that are not inherent to the concept of 'enhancement' itself. However, some motives for improvement carried through history are similar to those for enhancement.

Because of this ambivalence in defining enhancement, I will try to find another path for determining it:

1) A very common definition refers to the concept of normality: 'Human enhancement is the targeted technological enlargement or increase of human abilities over and above the "normal".'[38] However, discussions in social science, discourse theory, phenomenology, and social philosophy show that the term *normal* has many and incommensurable criteria of reference: for recommending enhancement – for instance, stimulating the growth of a child – it makes a difference whether we refer to *social* normality (i.e., an average health standard in a particular cultural context), or to *genetic* normality (i.e., with reference to biological heritage).[39] Thus, the concept of normality, seen as a zero point from which expansion or increase can be measured, seems highly ambiguous, and is therefore not to be recommended as the criterion for defining enhancement, or disability.

2) Another possible definition would be to say that enhancements are a 'new technology'[40] because they have 'much greater and more targeted enhancement effects than their predecessors' or any other therapeutic medicine.[41] Genetic diagnostic tools already have a great impact on prenatal diagnosis and decision-making; gene technology as a research or therapeutic possibility would fundamentally affect future generations. Thus, enhancement technologies sometimes have a strong quantitative increase in effect, but what is actually impressive is rather their new *quality*, a quality that will be discussed below. Hence, enhancement is a new technology because it affects the human condition and the design of the world in particular new ways. New technology can actually change structures and differences in society and culture, as well as the condition of the world; for instance, when assembly lines were introduced, humans learned to work in a particular rhythm: they had to adapt to the machine and they sometimes felt as if they were part of a big machine.

It is this second description that may lead to a deeper understanding of these new technologies that I call ordering or world-designing technologies. Today's new technologies, and certainly the fantasies that accompany them, have the capacity to invent a new order for some of the basic conditions of being human and thereby they design the condition of the world in a particular way that goes beyond an individual's decision or well-being.

## What could place the *conditio humana* at stake?

The notion of human condition refers to the world; more specifically, it refers to the question of how humans live in the world. Very importantly it tackles the meaning of being at home in the world or of being homeless, as Albert Camus brings out in his existential novels. The human condition encompasses the experiences of being human in a social, cultural, and personal context. However, the *conditio humana* has been associated with different dimensions of meaning; it can be interpreted as human nature or as human conditions.

Human nature refers to an essential determination of what it means to be a human. The support of such determination is often seen in natural laws which are supposed to govern the biological, physiological, or medical dimension of human beings in general. The sciences of biology, physiology, neurology, or other branches of medicine are certainly laden with cultural meaning. However, focusing on nature or on laws always means a specific scientific perspective, such as the world understood as environment, or the physiology of the body understood as a level of hormones or the functioning of muscles, medication, or nutrition that is adequate or inadequate for the body substance. Human nature is taken to be the fundamental nature and substance of humans.[42] However, as I have already hinted, 20th century debates are guided by skepticism about the determination of human nature and its moral use as an ontological basis for limiting enhancement.

The debate about whether human nature is determined or undetermined, open or closed, still focuses on the abstract concept of the human being in general and on its specific features or characteristics as a species. Thereby, the plurality of human beings is dwindled to the abstract singular.[43] Understanding 'condition' as nature implies the belief that we cannot escape it even though we have tried and continue to try to do so. Language fails when we try to define the human being: it fails because the *differentia specifica* of being human is founded only on the basis of the idea that a human is someone specific and singular. And this singularity cannot be captured in principle by a general definition. That is, the search for human nature seeks a general and substantial definition of the human being, yet this search reduces the human being to a general biological category by which humans no longer find their individual humanity.

Hannah Arendt clearly distinguishes human nature and the human condition: 'To avoid misunderstanding: the human condition is not the same as human nature, and the sum total of human activities and capabilities which correspond to the human condition does not constitute anything like human nature' (Arendt 1989, pp. 9–10). Human beings are conditioned and are creative in construing conditions. 'The human condition comprehends more than the conditions under which life has been given to man. Men are conditioned beings because everything they come in contact with turns immediately into a condition of their existence' (Arendt 1989, p. 9). In confirmation of Arendt's insight, the notion of human condition can be divided into two kinds of conditions, those that are discovered and those that are created. The discovered conditions are, for instance, mortality and natality, life, earth, worldliness (i.e., the dependency on objectivity), and plurality (i.e., fundamental differences between human beings). These aspects of human condition are grounding elements of experiences, and these experiences provide a basic foundation of reference for future experiences and judgments. Because humans have these constants of the experiential structure they can concretize the discovered conditions and thereby turn them into man-made conditions. Man-made conditions are for instance the cultural practices of how infants are born or how society treats their old people. The ability to transform conditions is essential to human life.

The history of philosophy has mostly been concerned with the human being in the singular, and in general; plurality was seen rather as a means for exclusion and, therefore, human relations, differences, heterogeneity, or contingencies could not be adequately grasped. This disregard of plurality is still found today and it remains a neglected problem of philosophical debates about anthropology or ethics.[44]

Since general and substantial definitions of human nature or the human condition must fail, they will never condition us absolutely, they will never tell us *what* we are or should be, and they can at best point to what Augustine mentioned when he stated, 'my God, give ear; look and see, [...], in whose sight I am become an enigma to myself; this itself is my weakness'.[45] And this can only be taken up and rephrased by the question 'who am I?' and by the stories we can tell. The notion of humans in the human condition allows for distinctions and plurality, and it opens up the way for an ethical dimension because it concerns relations, human relations, and how we live them. If humans are seen as being different and as having different looks, opinions, or perspectives, it becomes feasible to regard them as fundamentally relational subjects who are enabled to constitute relations with each other in better or worse, happier or unhappier ways.

In order to clarify the notion of the human condition, I will draw out some concrete features of this basic structure of our experiences. First, birth is the start of the human presence on earth and of the human relationship with the world around. Because men and women are born with the sense of having been begun by someone else, a woman, they are natal and have the capacity to begin. Furthermore, because of their birth humans are already situated in a base structure of generativity and relatedness.[46] Thus, when we construe the human condition from the perspective of being born (and not just of having to die), then it is a fact that a relation between at least two humans is the beginning. The concrete relationship (if good or bad, stable or broken) is apparent. Thus, the concreteness of this basic structure of being reveals not only the relatedness of generativity but also the principle of plurality among humans as a feature of the *conditio humana*.

Hence, second, 'not Man but men inhabit this planet. Plurality is the law of earth'.[47] The singular human does not make any sense for understanding ways of living and construing human relations. The plurality of ways of living and of relations of the lived worlds are always already construed and pre-given for everyone who is alive, but can also be changed and created by acting and intervening in the world. Third, life and striving for meaning are also human conditions. Such striving for meaning, its narrative and its strength, is historically, socially, and culturally (more or less) open in the future; constituting ways of life is a sense-giving procedure that may be more or less meaningful for the individual and her life. However, fourth, the future is not simply open, but from the individual perspective it is also finite. The sense of finitude depends upon mortality. Moreover, since humans have bodies, they are gendered, sensitive, and vulnerable. Vulnerability is a constant of experience, but its degree depends on the relationship to the world, personal situation, age, or strength. For instance, children are considered to belong to a 'vulnerable group', but how much they are confronted with their vulnerability depends upon the circumstances, their environment, and the care of other humans. Generally, vulnerability is the bodily feature that most affects the idea of deficiency or disability.

This list – natality, generativity, plurality, relatedness, earth, worldliness, mortality, temporality, spatiality, bodiliness, genderedness, sensitivity, and vulnerability – can certainly be expanded or thought of in a more detailed or precise way; the main point is that the notion of human conditions means an anthropological structure of constant experience. The concrete realization of the *conditio humana* supports, enforces, or limits the ways of living and acting in the world. Jürgen Habermas has pointed out the importance for human self-understanding of being 'grown' and of not having the feeling of having been 'made' by a scientist. He believes that only the gap between the beginning, over which a person does not have direct knowledge and over which she has no control, and the narrative exposition of her prehistory, provides the individual opportunity for a 'performative description of oneself, without which he or she would not understand him- or herself as the initiative person of his or her action or claims'.[48] Thus, the concept of human conditions is taken as a ground and a limit for human experience and acting in the world.

The foundation of the human conditions is associated with the fundamental base, possibility, and ability for change; yet taking it as a limit reflects the insight that our ways of acting are also 'limited by those things which men cannot change at will. And it is only by respecting its own borders that this realm, where we are free to act and to change, can remain intact, preserving its integrity and keeping its promises'.[49]

Human conditions seem to imply a necessary structure insofar as they refer to a grounded experience of being human, yet in terms of their appearance and the way we deal with them they are contingent foundations and contingent limits. Contingency, just like the notion of the human condition, encompasses both meanings: it provides the foundations of the ability to act but also sets limits on the capacity to act. This ambivalence means that the basic structures of human life are both disposable (not necessary, changeable) and non-disposable. In the latter sense they feel as if they are a result of personal fate, such as having a disability.

We can experience the contingency of the human condition through (at least) five different aspects:

- The materiality and functionality of the lived body, which are expressed in sensation and vulnerability.
- The finitude of life and the human possibility of choice of the 'appropriate form of life' for this lifespan.
- Strategies for human actions towards the surrounding world.
- Strategies for human self-empowerment, which presuppose the consciousness of freedom.
- Technological interventions into the modes of functioning of the body and of life, such as reproductive medicine and genetic or neurotechnologies.

The *conditio humana* constitutes the life of humans and their relationships. But since human conditions and features are either contingent, pre-given, or created, they can be brought out or suppressed in different ways and can lead to different social structures and relationships. It is the role of scientists, politicians, and philosophers to care for the fundamental aspects of the human condition in an appropriate way. Questions of what is appropriate for humans and for a particular society may never be answered absolutely, as new possibilities and opportunities lead to new questions. For instance, should it become the norm to control natality even more than it already is by using prenatal diagnosis and the subsequent selection of the embryo or fetus according to certain criteria of normality? How 'normal' should it be, for an individual or a society, for the procreation of a child to be totally distinct from sexuality; for instance, if the father is already dead when the mother conceives using his sperm? How would life and society be structured if humans lived forever, or if they lived for, say, 150 years? These are, technically speaking, almost simple questions; the hard question, however, is: how can the appreciation of plurality be furthered within society? Which technologies actually better such human relationships? How can we understand the entwinement of humans and machines? And what does it mean for the structure of justice and competition in sports, economics, or education? Would success in neuroscientific experiments with controlling the human mind lead to a happy life? Do we need such technologies in the military?

These and similar questions, if applied to different kinds of enhancement technologies, demonstrate that the application of enhancement technologies in a strong sense changes the order of the world, culture, society, and generations, and each person's life, decision-making, and responsibilities, in more or less drastic ways. Therefore, it would be appropriate to call enhancement technologies, which fundamentally restructure and order the world itself, 'design technologies'. And if a technology is not only designing an individual but essentially human relations, then this raises even more ethical questions.

## The worlds and the tasks of ethics

The technological, social, and cultural constitution of the human conditions determines how society, cultural practices, and ethical value systems are structured and practiced. For instance, the possibility of using cochlea implants as hearing aids means that the practice of using sign language is diminishing and disregarded. Vice versa, the design and the order of the world influence how the experiences of the human condition, such as plurality or generativity, can be practiced and lived by individuals. For instance, the availability of different reproductive technologies – from in-vitro fertilization through surrogate motherhood to pregnancy after menopause – pose a whole set of options which seem to make it absurd not to have a child. The observation is that the worldly conditions provide a certain realm of possibilities in which humans can decide, act, and experience; if sign language does not exist as means of communication then it is not part of the world any more (or only as dead language). If, for example, it is the practice to support a certain body growth then children will get growth hormones as a worldly standard. Already, many objects – from cars to kitchens – are designed for 'normally' grown people. These short examples do not imply that enhancement technologies are generally good or generally bad; rather they simply illustrate that the *conditio humana* corresponds to the *conditio mundana*. They correspond in such a way that a condition gets never changed alone. Hence, if enhancement technologies are found and implemented then this correspondence has to be kept in mind.

Ethics is concerned with ways of being in the world and modes of living with each other. The deep concerns of ethics are, hence, directed towards how we can or should live, understand, and experience the *conditio humana*. The basic question is thus: should we design and improve the world, society, and human relationships for humans and their relationships to each other, or should we design the individual human being, or future human beings, in order to fit them to society and adapt them to economic and technological needs? The discussion of enhancement eventually leads to the question of adaptation: must humans adapt to the capitalist world design and its technological provisions? Or can they resist the pressure put on them by such a system? From the standpoint of a capital system, humans are urged to adapt to the world which has been politically, normatively, technologically, and economically ruled and designed, structured and ordered, by a few men (and women).

The pressure for pregnant women to have several diagnostic checks seems to be premised on the idea of an individual decision, but it is accompanied by a whole set of questions imposed by a technological and medical industry. To then opt for prenatal genetic diagnostics means not only to support a certain industry; it also leads to a set of questions concerning economic and social considerations and allocated responsibilities. What should be done with genetic data? Who should receive these data – partners, family members, members of the coming generations, employers, insurance agents? Who will impose genetic testing, and on whom? I do not argue here that certain technologies are bad, rather I am indicating that the individual use of one technology inevitably structures a certain environment, furthers a particular human form, and supports certain ways of practice; likewise, the individual decision to use certain technologies is embedded in a whole set of 'professional' recommendations and promises, economic and social pressures. Particular technical practices are entwined with particular economic structures, and they impose their particular frame of questions on the individual.

Worldly conditions can be ordered in such a way that they are appropriate for certain kind of humans and not for others, as people with specific disabilities are well aware. Deaf people, for instance, without a hearing aid or sign language, may be simply unable to participate in discussions. Disability can be further understood using the disability theorist Rosemarie Garland-Thomson's (1997) concept of 'mis-fit' to describe the experience of persons who are physically impaired and therefore experience a 'mis-fit' between their bodies and their environments. Following up on this concept, Gail Weiss (2011) brings out the underlying unshared horizons and different meanings of the structures of daily life for those who find significance in them and those who are oppressed, politically disenfranchised, or physically excluded for having the 'wrong' body, 'wrong' race, or 'wrong' sexuality.[50]

The description of being a 'misfit' is a quite different statement. 'I don't fit' is a self-description or a description by someone else. Either way, it indicates a problem or a deviation, which hinders a person in regard to her basic needs or societal participation. The notion of misfit or non-misfit represents the hinge between the individual's condition and the worldly (mundane) condition in regard to space and time, social structure, and basic needs and desires.[51] The sense of being a non-misfit or misfit is often experienced as a basic condition of one's own possibilities, experiences, activities, or life in general, and as one criterion of the human condition in the world. Thus, the conditions of the world and the conditions of humans can be construed in such a way that many persons can participate in society, live meaningfully, or act comfortably. But the conditions of the world can also be designed in such a way that they initiate discriminatory behavior, exclude some humans, or force them to function only like machines.

Thus, it is the task of ethics to describe and evaluate the appropriateness of mundane conditions, material and structural relations, and orders of the world for human beings. Living humanely means, at least for most people, to have a world in which they can appear as individuals: distinctive from each other and non-interchangeable. Only a world of interaction, in which humans dwell and relate to each other,[52] provides a context for individual support and for human life, for plurality and for the constitution of a meaningful world in-between humans, who relate to one another.

If we argue that being human means inhabiting a man-made cultural and technical world, it is not self-evident what this means. As I have argued, striving for meaning is one of the human conditions, and we humanize the world by talking about it; thus, humans constitute a world as meaningful in-between themselves 'by acting and speaking'.[53] Things about which we cannot tell stories, which do not provoke narratives, lead not only to a form of dehumanization but also to a de-mundanization or unworldliness. Thus, when discussing 'enhancement', the criteria to reflect upon are not only the self-understandings of being human or the justice of allocating technological means, but rather evaluations of how the world is designed and structured for humans and their emotional, social, and political power relations. These are urgent questions concerning the order of the world and the possible threat of de-mundanization when the whole reference system of criteria changes and basic and stable structures of experiences given by the *conditio humana* and *conditio mundana* no longer seem certain. Thus, one might distinguish enhancement technologies which change certain humane and mundane conditions from radical enhancement technologies which focus on and threaten the foundational structure of the human condition itself. So far, I have focused on enhancement technologies that change human and mundane conditions but perhaps not in such a radical way. However, we can think about more drastic enhancements: for instance, if humans are not born from women anymore, then an important necessary basis of human self-understanding and feeling at home in the world would be turned to a contingent disposition.

Bernhard Waldenfels uses the term 'radical contingency' to refer to the idea of a collapse of the whole reference system.[54] Radical contingency implies that something can be different within a certain order and furthermore that the foundational structure itself is variable and disposable. This does not mean that the change takes place in the absence of any rules, but that the whole order, the structures of the apparently necessary human conditions – their material grounds and structural limits – become different. New technologies are new because they may have the power to endanger our reference system of criteria and our basic experiences. If humans were no longer born, if they were immortal, totally invulnerable,[55] or all 'look-alikes', then the world would become uncanny in a very radical way. In other words, the world would no longer be home; humans would feel homeless in a world that had become alien. The 'uncanny' calls for narratives to re-find the world in-between humans. Thus, our ethical project is to evaluate the radicalization of enhancement technologies and to consider the design and the order of the world in which humans feel at home.

I started this essay with some historical observations about the idea of improving deficient humans. A deficiency must be recognized in order to enhance something. Perhaps it is the most shocking insight that we are not healthy anymore, but permanently under the pressure of normalization and enhancement. Deficiency is in everybody and, therefore, everybody can and shall be improved. This results in an economy and morality to eradicate such deficiencies by enhancement technologies.[56]

## Notes

1. The term *conditio*, or condition, because it means conditionality or the constitution of the human, is best reflected by the German terms *Bedingtheit* and *Verfasstheit*. *Conditio* can be translated as condition, feature, but also as case of law, agreement, or contract. The *conditio humana* encompasses the different features of being human.
2. World Health Organization 2011, p. 3.
3. President's Council on Bioethics 2003, p. 17.
4. Parens 1998, p. 1. Parens strongly emphasizes the heterogeneity of enhancement technologies. This insight is important because not all have the same long-lasting or substantial impacts.
5. See for example: Fuchs et al. 2002, p. 16; Frankel and Kapustij 2008, pp. 55–58.
6. Müller 2010, pp. 157–159.
7. An interesting essay by Pöhlmann (1970, pp. 302–308) explores different views and metaphorical discourses about humans and their deficiencies.
8. President's Council on Bioethics 2003, p. 34.
9. Plato: Apologie, 21d. In order to find a perfect construction of the *polis* (state) Plato tells a 'noble fiction': the myth of the three metals – gold, silver, and bronze – which are mixed into human nature according to their status: gold for the king, silver for the guardians, and bronze for the craftsmen and peasants. Thus, human beings are prepared by birth and they have a certain nature, which can be improved according to their possibilities (Plato (1991): Politeia 414c–415d, III).
10. For Plato, birth is responsible for the soul's forgetting of all knowledge. This understanding is bound to three aspects: the entrance of the soul into life; the imprisonment of the soul in the body; and, hence, the transformation from a knowledgeable soul to a soul that does not know anything (see Schües 2008, pp. 35–39).
11. Plato: Apologie, 29e. Also, in the *Symposium* Plato (1991) develops a notion of *eros* and philosophy in which the idea of deficiency is central.
12. In Ancient Greek the term 'organism' is not seen as contrary to the non-living being, but has an instrumental character that in part lingers into the modern age (Meyer, 1969, p. 129). Plato already calls a part of the body *organon*, but it was Aristotle who made this term prominent in his biology. He calls *organon* all that is a corporeal means for the realization of the pre-given *eidos*, the anticipated form in the soul.
13. Accordingly the perfect state – the *polis* – is ordered by justice; that is, every man fulfills the role that is appropriate to him.
14. In several dialogues Plato refers to the ideal of the harmony of the bodily elements or the strings of mental and bodily senses (Timaeus (1925b) 24c, Phaedo (1966) 86c, Philebus (1925a) 47c).
15. Wiesing (2006, p. 324) suggests the concept of *restitutio ad integrarum*. However, this concept seems to mean computers or machines that function correctly or normally when one sets or resets the system; it does not capture the ancient idea of the good life.
16. Aristotle 1985. *Nicomachean Ethics* contains his reflections on the good life and aims to show how humans may live well.

17. Blumenberg 1981, p. 73.
18. Wiesing 2006, p. 325.
19. Röd 1995, p. 67.
20. See Arendt 1989, p. 311; the context of the notion of calculus and also the 'principle of happiness' is a discussion about utilitarian thinking; see also 154f.
21. However, this last point was debated in the 18th century, and there were understandings of humanity – such as those of Goethe or Humboldt – which acknowledged different cultures, religions, and traditions. This debate can be seen as one level of understanding humanity, as a general term, or in the sense of recognizing the individual.
22. The nature of the human is ethically, 'der subjektive Grund des Gebrauchs seiner Freiheit überhaupt (unter objektiven moralischen Gesetzen), der vor aller in die Sinne fallenden Tat vorhergeht' (Kant 1977b, p. 667).
23. 'Jedes Tier hat seinen Kreis, in den es von Geburt an gehört' (Herder (1772) 2001, p. 22).
24. Pöhlmann 1970.
25. Herder (1784–1791) 2002, p. 75 (my translation); 'ein ewiger Fortgang von organischer Schöpfung, der in jedes lebendige Geschöpf gelegt ward'. Kant himself also distinguished pragmatic and physiological anthropology.
26. Herder (1793–1797) 1968, p. 137.
27. Herder (1772) 2001, pp. 20, 24, 25. See also the article on 'Mängelwesen' (Brede 1998, pp. 712–713).
28. Herder (1772) 2001, p. 22 (my translation); 'so enge und einförmige Sphäre'.
29. Herder (1784–1791) 2002, p. 135.
30. Nietzsche 1988, p. 81.
31. Gehlen 1986, p. 36, also pp. 41, 51 (*Reizüberflutung*).
32. Gehlen 1986, pp. 38, 348.
33. Gehlen invented – because human beings are weak and open and exposed to the world – his rather controversial doctrine of institutions and higher leadership systems (*oberste Führungssysteme*), which played into the hands of National Socialism. He raised the notion of a deficient being from a methodological device to a manifestation of the threat of a human being who can degenerate, and therefore needs rigid leadership systems (Gehlen 1956, p. 18). Max Scheler pushes Gehlen's theory further by characterizing the mind as the hypertrophy of a being with brain, which is organically a faux pas of nature, a blind alley of life (Scheler 1976, p. 139).
34. Plessner 2003, p. 15.
35. Galton 1973, p. 17. Genetic theory began here.
36. Ehrenberg 2008, pp. 20, 14.
37. As discussed by Bosteels and Blume in Chapter 5 of this volume.
38. Hennen 2008, p. 97. Concerning the notion of 'normality' and about the so-called 'normal function model' see Parens 1998, p. 3. See also the essay by Scully and Rehmann-Sutter 2001, who argue strongly against a normality-based enhancement definition. This argument is also described in the interview with Jackie Leach Scully in Chapter 2.
39. This thought refers to a rather well-known short story about two 11-year-old boys who both have the prognosis of growing to a height of only 160 cm (Parens 1998, p. 6).

40. The term 'new technology' is mentioned in almost all publications when it comes to blurred boundaries between restorative or preventive, non-enhancing therapy, therapeutic enhancements, or non-therapeutic enhancement. Stoa 2009, p. 6.
41. Mehlmann 2009, p. 2.
42. The term 'human nature' has also been determined by religious beliefs, psychological studies, or different kind of ideologies. Also these positions work with the idea of a human substance and essence.
43. One of the most well-known debates was led by Noam Chomsky and Michel Foucault 1971 (Chomsky and Foucault 2006); since then the question of whether human nature should be taken as an ontological basis or rather as culturally constituted has been extensively discussed. See Weiß 2009.
44. Even though postmodernists and poststructuralists try to disperse the singular of the human being with notions of difference or contingency, thinking in terms of a plurality of humans – without treating everything merely as equivalents and still keeping an evaluative standard – remains a challenge. The recognition of the fundamental plurality among humans is essential to respect gender or cultural differences.
45. Augustine 10.33.50.
46. Schües 2008, pp. 323–396.
47. Arendt 1978, p. 19.
48. Habermas 2005, p. 103.
49. Arendt 1987, pp. 263–264.
50. Weiß 2011, p. 173.
51. The term non-misfit is intentionally used here. Any positive term, such as normal or fit, implies an exclusionary usage.
52. With reference to spatial and interrelational interpretations of the concept of dwelling in the world, Heidegger (1952) is inspiring; see also Schües 2012.
53. Arendt 1989, ch. 5.
54. Waldenfels 1998, p. 53.
55. McKenny (1998, pp. 222–237) focuses on the central notion of vulnerability.
56. I thank Zoe Goldstein, Christoph Rehmann-Sutter, and Rudolf Rehn for valuable questions and comments.

## References


Arendt, H. (1978): *Life of the Mind*. San Diego: Harcourt Brace Jovanovich.

Arendt, H. (1987): Truth and politics. In: *Between Past and Future*, 11th edn. New York: Penguin Books, pp. 227–264.

Arendt, H. (1989): *The Human Condition*. Chicago: University of Chicago Press.

Aristotle (1985): *Nicomachean Ethics*, translated by Irwin, T. Indianapolis, Indiana: Hackett Publishing.

Augustine, Stant: *Confessiones*, translated and edited by Outler, A. C. http://www.naderlibrary.com/cult.confessionsaugustine.10.html, chapter 33.

Blumenberg, H. (1981): *Wirklichkeiten, in denen wir leben*. Stuttgart: Reclam.

Brede, W.(1998): Mängelwesen. In: Ritter, J. (ed.) *Historisches Wörterbuch der Philosophie*. vol 5, Basel: Schwabe, pp. 712–713.

Chomsky, N. and Foucault, M. (2006): *The Chomsky–Foucault Debate On Human Nature*. New York: The New Press.

Ehrenberg, A. (2008): *Das erschöpfte Selbst. Depression und Gesellschaft in der Gegenwart*. Frankfurt/M.: Suhrkamp.
Epicurus (2011): Letter to Menoeceus, translated by Saint-Andre, P. http://www.monadnock.net/epicurus/letter.html.
Frankel, M. S. and Kapustij, C. J. (2008): Enhancing humans. In: Crowley, M. (ed.) *From Birth to Death and Bench to Clinic: The Hastings Center Bioethics Briefing Book for Journalists, Policymakers, and Campaigns*. Garrisson: The Hastings Center, pp. 55–58.
Fuchs, M., Lanzerath, D., Hillebrand, I., Runkel, T., Balkerak, M. and Schmitz, B. (2002): Enhancement. Die ethische Diskussion über biomedizinische Verbesserungen des Menschen. In: *drze-Sachstandsbericht*, vol 1 Bonn: Deutsches Referenzzentrum für Ethik in den Biowissenschaften.
Galton, F. (1973): *Inquiries Into Human Faculty and Its Development*, 2nd edn. New York: AMS Press.
Gehlen, A. (1956): *Urmensch und Spätkultur. Philosophische Ergebnisse und Aussagen*. Bonn: Athenäum.
Gehlen, A. (1986): *Der Mensch. Seine Natur und seine Stellung in der Welt*, 13th edn. Wiesbaden: Aula-Verlag.
Habermas, J. (2005): *Die Zukunft der menschlichen Natur. Auf dem Weg zu einer liberalen Eugenik?*. Frankfurt/M.: Suhrkamp.
Heidegger, M. (1952): 'Bauen, Wohnen, Denken.' In: *Vorträge und Aufsätze*. (Pfullingen: Neske 1985), pp. 139–156.
Hennen, L. (2008): Review of Zonneveld L., Dijstelbloem H., Ringoir D. Reshaping the human condition. Exploring human enhancement. In: *Technikfolgenabschätzung – Theorie und Praxis*, Institut für Technikfolgenabschätzung und Systemanalyse (ITAS), No. 1, 19th year – April 2010, pp. 97–101, http://www.itas.fzk.de/tatup/101/henn10a.htm.
Herder, J. G. (1968): Briefe zur Humanität (1793–1797). In: Menze, C. (ed.) *Johann Gottfried Herder. Humanität und Erziehung*, 2nd edn. Paderborn: Schöning, pp. 136–141.
Herder, J. G. (2001): *Abhandlung über den Ursprung der Sprache (1772)*. Stuttgart: Reclam.
Herder, J. G. (2002): Ideen zur Philosophie der Geschichte der Menschheit 1784–1791. In: Pross, W. (ed.) *Ideen zur Philosophie der Geschichte der Menschheit*, vol III/1 München: Hanser, pp. 95–135.
Kant, I. (1977a): Anthropologie in pragmatischer Hinsicht. In: Weischedel, W. (ed.) *Schriften zur Anthropologie, Geschichtsphilosophie, Politik und Pädagogik 2*, Werkausgabe vol XII Frankfurt/M.: Suhrkamp, pp. 399–690.
Kant, I. (1977b): Die Religion innerhalb der Grenzen der bloßen Vernunft. In: Weischedel, W. (ed.) *Schriften zur Anthropologie, Geschichtsphilosophie, Politik und Pädagogik 2*, vol VIII Frankfurt/M.: Suhrkamp, pp. 648–879.
La Mettrie, J. O. de (1999): *L'Homme Machine (1748)*. Paris: Flammarion.
McKenny, G. P. (1998): Enhancement and the ethical significance of vulnerability. In: Parens, E. (ed.) *Enhancing Human Traits: Ethical and Social Implications*. Washington: Georgetown University Press, pp. 222–237.
Mehlmann, M. (2009): *The Price of Perfection. Individualism and Society in the Era of Biomedical Enhancement*. Baltimore: The John Hopkins University Press.
Meyer, A. (1969): Mechanische und organische Metaphorik politischer Philosophie. In: *Archiv für Begriffsgeschichte*, 13, pp. 128–199.

Müller, O. (2010): *Zwischen Mensch und Maschine. Vom Glück und Unglück des Homo faber.* Frankfurt/M.: Suhrkamp.

Nietzsche, F. (2010): Jenseits von Gut und Böse. In: Colli, G. and Montinari, M. (eds.) *Kritische Studienausgabe,* vol V Berlin: de Gruyter 2010.

Parens, E. (1998): Is better always good? The enhancement project. In: Parens, E. (ed.) *Enhancing Human Traits. Ethical and Social Implications.* Washington: Georgetown University Press, pp. 1–28.

Plato (1925a): Philebus. In: *Plato in Twelve Volumes,* translated by Fowler, H. N. vol 9. Cambridge: Harvard University Press; London: William Heinemann Ltd.

Plato (1925b): Timaeus. In: *Plato in Twelve Volumes,* translated by Lamb, W. R. M. vol 9 Cambridge: Harvard University Press; London: William Heinemann Ltd.

Plato (1966): Apologie. In: *Plato in Twelve Volumes,* translated by Fowler, H. N. introduction by Lamb, W. R. M. vol 1 Cambridge: Harvard University Press; London: William Heinemann Ltd.

Plato (1966): Phaedo. In: *Plato in Twelve Volumes,* translated by Fowler, H. N. introduction by Lamb, W. R. M. vol 1. Cambridge: Harvard University Press; London: William Heinemann Ltd.

Plato (1967): Protagoras. In: *Plato in Twelve Volumes,* translated by Lamb, W. R. M. vol 3. Cambridge: Harvard University Press; London: William Heinemann Ltd.

Plato (1991): Politeia. In: *The Republic of Plato,* translated by Bloom, A. 2nd edn. New York: Basic Books.

Plato (1991): *Symposium,* translated and commented by Allen, R. E. New Heaven: Yale University Press.

Plessner, H. (2003): Die Frage nach der Conditio humana (1961). In: *Gesammelte Schriften VIII.* Darmstadt: Wissenschaftliche Buchgesellschaft, pp. 136–217.

Pöhlmann, E. (1970): Der Mensch – das Mängelwesen? Zum Nachwirken antiker Anthropologie bei Arnold Gehlen. In: *Archiv für Kulturgeschichte,* 52, pp. 297–312.

President's Council on Bioethics (2003): *Beyond Therapy. Biotechnology and the Pursuit of Happiness. A Report of the President's Council on Bioethics.* New York: Dana Press.

Röd, W. (1995): *Descartes. Die Genese des cartesianischen Rationalismus.* 3rd edn. München: Beck.

Scheler, M. (1976): Mensch und Geschichte. In: Frings, M. S. (ed.) *Späte Schriften,* vol 9 Bern: Bouvier.

Schües, C. (2008): *Philosophie des Geborenseins.* Freiburg/Br.: Alber.

Schües, C. (2012): Nachbarschaft – Eine fragile Beziehung. In Staudigl, M. (ed.) *Phänomenologien der Gewalt.* Paderborn: Fink 2014, pp. 333–351.

Scully, J. L. and Rehmann-Sutter, C. (2001): When norms normalize: The case of genetic 'Enhancement.' In: *Human Gene Therapy,* 12(1), pp. 87–95.

STOA (Science and Technology Options Assessment), European Parliament (2009): Human enhancement study, http://www.itas.fzk.de/deu/lit/2009/coua09a.pdf.

Thomson, R. G. (1997): *Extraordinary Bodies: Figuring Physical Disability in American Culture and Literature.* New York: Columbia University Press.

Waldenfels, B. (1998): *Grenzen der Normalisierung.* Frankfurt/M.: Suhrkamp.

Weiß, M. G. (2009): *Bios and Zoë. Die menschliche Natur im Zeitalter ihrer technischen Reproduzierbarkeit*. Frankfurt/M: Suhrkamp.
Weiss, G. (2011): Sharing time across unshared horizons. In: Schües, C. Olkowski, D. Fielding, H. A. (eds.) *Feminist Phenomenology*. Bloomington: Indiana University Press, pp. 171–189.
Wiesing, U. (2006): Zur Geschichte der Verbesserung des Menschen. Von der restitutio ad integrum zur transformatio ad optimum?. In: *Zeitschrift für medizinische Ethik*, 52, pp. 323–338.
World Health Organization (2011): *World Report on Disability*, http://www.who.int/disabilities/world_report/2011/en/index.html.

# Part II

# Case Studies

# 4

# Good Old Brains: How Concerns About an Aging Society and Ideas About Cognitive Enhancement Interact in Neuroscience

*Morten H. Bülow*

Discussions about human cognitive enhancement are based, in various ways, upon assumptions about neuroscientific knowledge production and the applicability of neuroscientific results. From utopian imaginings of (trans)human ascension to current practices of using pharmaceuticals, developed for different neurodegenerative conditions, as so-called 'smart drugs', these assumptions all seem to revolve around the expectations of and hopes for what neuroscience can or will be able to do. It is striking, but maybe not surprising, that discourses and practices surrounding cognitive enhancement have flourished in the same historical period that has seen a 'neuroscientific turn' in the natural sciences (Littlefield and Johnson 2012) and widespread popularity of the neuro-prefix in many different social and cultural settings (Frazetto and Anker 2009). It has become almost trivial to make the general observation that developments in neuroscience do not take place in a social or cultural vacuum – that they both affect and are affected by the context in which they are situated – and that the debates and practices related to cognitive enhancement (including this chapter) can be seen as part of this complex interrelation between science and society. It might be more interesting to move one step closer and ask what it is in the neurosciences that relates to discussions about human enhancement, or, more specifically, how the production of knowledge within the neurosciences relates to hopes and wishes for cognitive enhancement.

Another point of departure for this chapter is that enhancement debates have a strong, often overlooked, connection to developments in the field of aging research and related discourses about aging in society. Like enhancement debates and neuroscience, concepts such as 'successful aging' became widespread from the 1980s onwards, forming a 'new gerontology' (Rowe and Kahn 1998), which has become paradigmatic for developments in aging research. Successful aging is a conceptual framework generated (among other things) by a rising concern for the possible consequences of aging populations; and through countering this concern, with emphasis on the positive aspects and possible optimization of the aging process, it places itself at the intersection of the scientific and social concerns that drive fundamental parts of both neuroscience and enhancement debates. An investigation of the neuroscientific practices related to such notions as 'successful aging' may therefore also offer new perspectives on the human enhancement debate.

Another important element in this story is that this intersection between neuroscientific practices, notions of successful aging, and enhancement debates very much relates to definitions of normality and what it means to be human, and that the notions of and practices related to successful aging, human enhancement, and neuroscience seem to contain a shift in perceptions of normality. For example, it is argued repeatedly in connection to neuroscience in general that knowledge about the brain has the potential to change understandings of what it means to be human; within aging research there is constant debate about the definition of 'normal', 'pathological', and 'successful' aging; age-related neurodegenerative conditions such as Alzheimer's disease are said to be 'dehumanizing'; and debates about human enhancement sometimes refer to notions such as 'transhumanism' or 'the posthuman'. But how are all these intersecting phenomena related, and how do they change notions of normality?

Drawing on a historical analysis of the concept of successful aging in neuroscience publications from the 1980s to today, this chapter will discuss how the aims and the production of knowledge within age-related neuroscience are connected to ideas about cognitive enhancement. One specific discussion point will be the concept of normality and the changes in our conception of what it means to be human that might be illuminated by investigating these interconnected fields. I will first introduce the notion of successful aging and point to the central claims about normality, causality, and the purpose of aging research it contains. Then, I will relate this conceptual framework of successful aging to current discussions about human enhancement, and illustrate how these ideas do (or do not) interconnect in neuroscience.

## Successful aging

In the context of aging research, the notion of successful aging can be found as far back as the 1960s, but it first became widespread in the late 1980s after gerontologist John Rowe and psychologist Robert Kahn published an article in *Science* in 1987, entitled Human aging: Usual and successful (Rowe and Kahn 1987). Rowe and Kahn's article has since been widely cited in scientific publications in a variety of disciplines, from molecular science and odontology to psychology and sociology.

The article can be characterized as seeking to change the focus, methods, and understandings of aging within aging research, challenging the fundamental concepts of what constitutes aging and what constitutes disease. The question of what aging 'is' – the ontology of aging – also quickly becomes a question of what constitutes (or should constitute) a 'normal' human life course; a question that is inseparable from issues of how researchers should gain knowledge of aging and practice science, as well as what kinds of clinical or social interventions researchers suggest, imagine, or practice. It is not a trivial question and, as previously mentioned, it is also fundamental to the discussions about human enhancement to which I return later.

## Challenging 'normality'

In Rowe and Kahn's perspective, in 1987 when they wrote their paper the idea of normality in aging research had previously sustained three central problems. First, it divided people into categories of 'normal' versus 'diseased', and thereby neglected the vast heterogeneity among older people. Second, it had applied the concept of 'normal' to a large group of people, implying that they were in a harmless or risk-free situation. And third, using the concept of 'normal' somehow implied that the condition of normality was also 'natural' and therefore should not be modified. In other words, according to Rowe and Kahn, this notion of normality had generalized and naturalized the conditions of a large group of older people, placing them outside of the concern of medical studies on aging.

On the whole, Rowe and Kahn further argued, this concept of normality tended to create 'a gerontology of the usual': by neglecting the large group of people placed within the broad category of normality, gerontologists were in actuality promoting 'usualness' instead of what might be considered as 'success' (Rowe and Kahn 1987, p. 143). The problem with this, in Rowe and Kahn's view, was that 'the usual' usually led to functional decline or disease later in life – the 'usual' would on average thus only be a precursor of disease and decline. In this way, the usual did not really seem to be normal but rather potentially, or even pre-symptomatically, ill – or at risk.

Instead, Rowe and Kahn suggested that 'normal aging' should be viewed as consisting of both 'usual aging' and 'successful aging'. Success is possible, they argued, because of the existence of 'older persons with minimal physiologic loss, or none at all, when compared to the average of their younger counterparts' (Rowe and Kahn 1987: pp. 143–144). Successfully aging individuals, then, are defined as individuals who have little or no loss of physiological functions, which makes the achievement of such a 'successful' condition the objective for aging research (Rowe and Kahn 1987, p. 144).

## Challenging 'causality'

Whereas the 'gerontology of the usual' considered average age-related decline as something 'age-intrinsic' (Ibid.), Rowe and Kahn's new perspective shifted consideration towards a regard of the average age-related decline as something very much dependent on social and environmental factors:

> It is at least a reasonable hypothesis [...] that attributions of change to age per se may often be exaggerated and that factors of diet, exercise, nutrition, and the like may have been underestimated or ignored as potential moderators of the aging process. If so, the prospects for avoidance or even reversal of functional loss with age are vastly improved, and thus the risk of adverse health outcomes reduced.
>
> (Rowe and Kahn 1987, p. 144)

In opposition to a view of aging as something defined by inevitable physiological and cognitive decline, the notion of successful aging stresses that individuals can avoid such decline by maintaining and improving themselves through changeable lifestyle factors. In essence, this view emphasizes the aging process as indeterminate, open-ended, and multi-causal.

This view of aging also entails a shift in research perspective from a focus on treating disease to preventing disease – and preventing the factors related to 'usual' aging that eventually lead to disease or decline in physical or mental functioning. Pushing this interpretation a bit further, it does not seem farfetched to suggest that underlying this shift in research focus is the view that successful aging is the normal state that the body should – ideally – be in. Since everybody who, by Rowe and Kahn's definition, does not belong to the category seems to be in need of some sort of medical intervention or other, Rowe and Kahn can be seen to refute and pathologize the notion of the 'usual' as representative of aging per se, and instead to promote the notion of an able, disease-free body as a successful and normal body. When related to the enhancement debate, this means that being (feeling?) averagely well does not seem to be good enough. To paraphrase Elliot (2003), you have to be better than well to age successfully. I will come back to this shortly.

## Enhancement and successful aging

The notion and aims of successful aging fit well with the discourses of public health that have developed in most Western countries since the 1980s, where individualized responsibility for healthcare (or perhaps self-care) in the form of fitness and healthy lifestyles have increasingly been promoted (Petersen 1996), and medical technologies not only seek to treat and correct but also to 'optimize' individuals (Rose 2007, cf. Petersen and Wilkinson 2008). Furthermore, as mentioned earlier, the notion of successful aging (and related notions such as optimal aging, healthy aging, positive aging, and so on) can be seen as part of a discourse of growing concern from the 1980s onwards about the aging populations of Western societies. Consider the large number of potentially, or pre-symptomatically, ill 'usual agers' in Rowe and Kahn's definition. Framed in such a way, it becomes difficult to see them as anything but a ticking time bomb within the healthcare system, an age-related disease and disability catastrophe waiting to happen – unless they are somehow physically and mentally transferred to the category of 'successful agers' (cf. Kampf and Botelho 2009; for a critical gerontological analysis of the concept of successful aging, see for instance Cole 1992, p. 238 and Holstein and Minkler 2003).

The concerns related to the notion of an aging society – and the consequences for national healthcare systems, pension funds, the maintenance of a healthy workforce, and the ability to compete in a globalized market – have provided motivation for promoting the individualized solutions (and hope) offered by notions like 'successful aging'; and perhaps also for promoting human enhancement. Both the European Union (EU) and the United States National Science Foundation have published reports dealing with so-called 'converging technologies' (CT), in which human enhancement is a central point of discussion and is seen to relate to aging research (Roco and Bainbridge 2002; Giorgi and Luce 2007). As a study commissioned by the EU parliament states, 'it is safe to say that a side effect of the fast-growing research and development into pharmaceuticals for age-related neurodegenerative diseases will be a number of new drugs which can be used for the enhancement of performance of young, healthy people' (Human Enhancement Study 2009, p. 7).

Similarly, in enhancement debates on websites belonging to groups of individuals who support and promote the idea of human enhancement (e.g., the transhumanist movement), one of the central hopes and arguments for developing enhancement technologies is the possibility of using them as life extension devices to ward off age-related physical or cognitive decline (Bostrom 2005; Humanity+). Both pro-enhancement advocates and EU reports stress that biomedical science is increasingly producing technologies that have the potential to enhance the capacities of healthy people, as well as to treat disease (Giorgi and Luce 2007; Human Enhancement Study 2009; Humanity+).

Neuroscience is recognized as one of the dominant biomedical sciences of interest for both aging research and enhancement debates (Giorgi and Luce 2007; Human Enhancement Study 2009; Kirk 2008; Mehlman 2004). Not only are the brain and cognition ascribed significant cultural value in today's Western ('knowledge') societies, but developments within neuroscientific aging research also seem to have significant connections to cognitive enhancement: 'The growing problem of neurodegenerative diseases in aging societies has turned research and development in therapeutic cognitive enhancers into a very dynamic field with significant resources' (Human Enhancement Study 2009, p. 26; see also Daffner 2010). Both individuals and societies, it seems, have reason to improve cognitive functions and prevent neurodegenerative diseases (cf. Beddington et al. 2008).

Furthermore, as mentioned in the introduction, ideas and knowledge about cognition and brain functions have a major impact both in the cultural sphere and on people's views of self and identity (Dumit 2004; Frazetto and Anker 2009; Rose 2003). As one study related to the discussion of CT points out:

> 'neuro/brain enhancement' as a research field stands at the centre of the CT debate. It attracts the largest share of attention due to its plans to simulate and manipulate brain processes, which – if realized successfully – could directly affect our concepts of the human self and identity.
>
> (Beckert, Blümel and Friedewald 2007, p. 382)

These are not just abstract thoughts about possible futures. Medications developed through research into age-related conditions such as Alzheimer's disease are already being used by young, healthy individuals to enhance their cognitive abilities (Beckert, Blümel and Friedewald 2007; Greely et al. 2008; Human Enhancement Study 2009; Humle and Friislund 2010). Not only is there a kind of underlying demand for optimization and enhancement in the present discourse about aging, but one possible (spreading) side effect of the huge focus on conditions such as Alzheimer's disease is the production of new drugs that can be used for other unintended purposes, such as cognitive enhancement of young and healthy people. Medication for the old turns into enhancement for the young.

## Neuroscience

So how do these debates and emerging practices relate to the production of knowledge within neuroscience, in particular neuroscience related to aging research? To approach this question let us again look at the notions of normality and enhancement, this time focusing on neuroscientific publications.

Many of the ever-increasing number of age-related neuroscientific publications of the last 20 years or so deal with different aspects of age-related cognitive impairment, predominantly (but not solely) Alzheimer's disease. A major (implicit or explicit) point for discussion in these publications is therefore – as in Rowe and Kahn's article on successful aging – what to define as part of a normal aging process and what should be considered as pathological aging. The publications also seem to share the common goal of producing knowledge that will somehow, potentially, lead to interventions for conditions of impairment (be they normal or pathological). As a review article in Current Opinion in Neurobiology states:

> New discoveries challenge the long-held view that aging is characterized by progressive loss and decline. Evidence for functional reorganization, compensation and effective interventions holds promise for a more optimistic view of neurocognitive status in later life. Complexities associated with assigning function to age-specific activation patterns must be considered relative to performance and in light of pathological aging. New biological and genetic markers, coupled with advances in imaging technologies, are enabling more precise characterization of healthy aging. This interdisciplinary, cognitive neuroscience approach reveals dynamic and optimizing processes

in aging that might be harnessed to foster the successful aging of the mind.

(Reuter-Lorenz and Lustig 2005, p. 245)

This is just one example of many within the field. All in all, neuroscientific research has facilitated a view of brain functions that emphasizes on the one hand plasticity and complexity, and on the other the search for biological markers and pathways to allow for interventions, with the twofold aim of treating conditions such as Alzheimer's disease (mentioned later in the same article) as well as optimizing the aging mind (see also Daffner 2010). Genes, cardiovascular function, cognitive plasticity and the ability to compensate, cognitive training, emotional bias, and more, are all given as factors involved in the performance and processes of the brain (Reuter-Lorenz and Lustig 2005, p. 247). This complexity makes identifying 'normal' or 'healthy' characteristics difficult, since every individual life course is unique and has a unique combination of factors affecting the constitution of the brain (this is even so, one might add, for laboratory rats).

Even in the case of a well-known neurodegenerative disease such as Alzheimer's, there has been – and still is – considerable debate about how to characterize the condition (Lock 2005; see also Whitehouse and George 2008). Although Alzheimer's disease has some defining characteristics, such as brain 'plaques' and 'tangles', no necessary correlation has been found between the quantity of such brain 'anomalies' and the cognitive functioning of an individual; that is, you can have a biologically 'abnormal' brain and still function well, or a not-so-abnormal brain and have severe problems (Rose 2009). In practice, the diagnosis of Alzheimer's disease can only be 'probable' since multiple factors can contribute to many of the symptoms, and the clinical diagnosis can (so far) only be validated post mortem. Perhaps because of this uncertainty and complexity, one question that keeps recurring in the research has been how to separate normal from pathological brain aging. Is what is seen as normal brain aging in itself a sign of dementia, that is, something that is not normal but diseased; or conversely, are dementias such as Alzheimer's disease actually a normal development for the aging brain (cf. Goodwin 1991)?

Returning to the topic of enhancement, when discussing how to avoid, prevent, or postpone neurodegenerative diseases such as Alzheimer's, one of the concepts used to explain the possible optimization of and positive intervention into age-related brain processes is that of 'cognitive reserve' (Daffner 2010; Liberati, Raffone and Belardinelli

2011). As the name implies, the idea behind cognitive reserve is that if a person builds up mental capacities at an early age, through a combination of proper nutrition, education, and cultural stimulation, the effects of this build-up may function as a reserve to be utilized at later stages of life. The accumulation of cognitive reserves is theorized to work either through providing a strong cognitive system with built-in resilience, or by providing a plethora of resources enabling compensation for age-related cognitive difficulties or decline (Daffner 2010; Kirk 2008; Liberati, Raffone and Belardinelli 2011).

Building up cognitive reserves may even be seen as an attempt to raise the threshold before the inevitable commencement of functional decline reaches it with time. Raising the threshold, so to speak, means applying neuroscientific knowledge to utilize the body's own potential to enhance and protect cognitive functions that would otherwise – usually – have been in a weaker position and would have suffered decline sooner and/or more severely. Consequently, striving for more cognitive reserves could also be seen as striving towards raising the bar of normality, of what 'ordinary' human cognition throughout life and in old age should be.

## Perspectives on enhancement

The language of age-related neuroscience is not far from that of enhancement debates. After all, neuroscientific articles have for the last 20 years often written conclusions or abstracts similar to Collier and Coleman, who state that their 'findings suggest ways in which biological aging can be manipulated to promote good function in aged individuals' (Collier and Coleman 1991, p. 685). This seems to offer exactly what enhancement discussions are about: promising biomedical or technological developments that can promote 'good function'. However, while neuroscientific aging research might suggest future optimizing interventions or implicit ways of changing what could be viewed as normal aging – as per enhancement debates – it also offers a somewhat different perspective. What most neuroscientific developments imply is not the improvement of functioning beyond the 'normal', but rather a shift in focus from one that aims at the average within the range of normal functions, to one that aims at optimization within the normal range. And more often than not, quotes such as Collier and Coleman's above simply refer to treating functional decline that has already taken place – not preventing it from happening in the first place.

Nevertheless, both neuroscience in general and research on successful aging in particular focus on preventive measures, and these preventive measures often operate by improving the cognitive function of otherwise 'normal' individuals through lifestyle interventions or substance intake. A prominent discussion within enhancement debates concerns the issue of when a medical, technological, or pharmaceutical intervention should be viewed as a treatment and when it counts as enhancement. But I would suggest that this distinction may not be the most difficult or fundamental one. Therapy implies that the subject is in a condition where she/he is deemed to function less well than is thought appropriate for the situation (e.g., a child who cannot hear might be thought to be in need of a cochlear implant), while enhancement implies an improvement of something that is already functioning well, though not excellently (e.g., golf player Tiger Woods' vision-improving operation). Although the judgment of whether something functions well or less well than appropriate is of course arbitrary and dependent on negotiable definitions of normality, this is still a distinction one can use as an analytic tool to clarify whether something is viewed as an enhancement in a specific historical and cultural context. A much more difficult distinction, in my opinion, is the one between prevention and enhancement.

As mentioned earlier, preventive measures often work by improving the cognitive function of otherwise 'normal' individuals, something that can be viewed as closely related to issues of optimization. To borrow the words of British sociologist Nikolas Rose: 'what is involved here cannot be divided according to the binary logic of treatment versus enhancement; it is a constant work of modulation of the self in relation to desired forms of life' (Rose 2009, p. 80). In the context of this chapter we could also ask: is optimization of human bodily potential an enhancement or is it still within the range of 'normality'?

## Some concluding remarks

A close relationship can be found between certain aspects of the enhancement debate and the knowledge production disseminated through neuroscientific publications about successful aging: both seek the improvement or optimization of bodily and/or cognitive functions and a long and healthy life, and both focus on the development of new medical or technological methods to facilitate these measures. Both the notions of successful aging and enhancement are centered on individual capabilities, and neither (from the outset) take much notice of social inequalities. There is, furthermore, a certain orientation towards the future in neuroscience, successful aging, and enhancement: attaining the best possible future is what cognitive reserve, prevention, and improvement of human capacities are all about.

But there is also a discrepancy between issues of prevention/optimiza tion and enhancement – neuroscience and successful aging are not posthuman or transhuman, although this does not mean that neither field can transform our understanding of the human condition. The last couple of decades have seen a change in expectations of particular scientific fields such as neuroscience or genetics, and of research and discourses concerning aging; but also perhaps of more fundamental concepts concerning normality and causality. In some ways normality has been pathologized, or, as I would suggest here, pushed to include a more demanding regime of prevention and optimization. Underlying these concepts – and contained within the changes in scientific perspectives – new ideas about what it means to be human seem to be emerging. We can see the practical and political consequences of these new ontologies reflected in the studies of human enhancement, but perhaps it is also time to delve into their foundations.

Let me end this chapter on a slightly different note. In 2002 the French artist Gilles Barbier exhibited an installation of six wax models of life-sized superheroes, shown in the picture below (Figure 4.1). Aging and enhancement seem to be the key topics for this clever piece, which

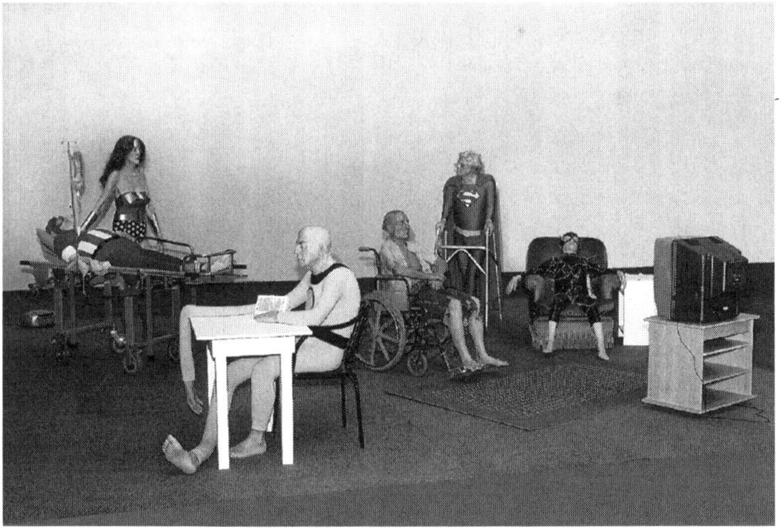

*Figure 4.1* Gilles Barbier de Preville: L'hospice. © VG Bild-Kunst, Bonn 2013

somehow functions as an illustration of both the hopes and anxieties involved. Here we can laugh at the witty materialization of aging superheroes, and we might even laugh out of relief because we know that they are fiction. Maybe the laughter even has a cathartic effect, reminding the viewer of his/her own mortality, of time passing, and of the seemingly inevitable age-related decline in physical and mental functioning. Who has not wanted to be a timeless, never aging, physically and/or mentally enhanced superhero?

Barbier's installation can also be seen as a comment on the current discussions and concerns about aging in most Western countries, upon which this chapter has touched. High functioning individuals – embodied here as 'the superhero' – age successfully; they do not end up sitting in a wheelchair in a care home. So this artwork seems to point not only to the ideals of popular culture, but also to contemporary popular discourses on aging and debates about human enhancement. Like a joke, it turns our expectations upside down. But the installation also counterposes this joke with a more serious question, one that the American historian of aging, Thomas Cole, posed at an aging conference in Copenhagen in 2011, and that is pertinent considering the discursive and ontological transformations I have outlined here: what value do we place on frail people within society?

## References

Beckert, B., Blümel, C. and Friedewald, M. (2007): Visions and realities in converging technologies. In: Giorgi, L. and Luce, J. (eds.) *Converging Science and Technologies: Research Trajectories and Institutional Settings*, special issue in *Innovation. The European Journal of Social Science Research*, 20(4), pp. 375–395.

Beddington, J., Cooper, C. L., Field, J., Goswami, U., Huppert, F. A., Jenkins, R., Jones, H. S., Kirkwood, T. B. L., Sahakian, B. J. and Thomas, S. M. (2008): The mental wealth of nations. In: *Nature*, 455, pp. 1057–1060.

Bostrom, N. (2005): A history of transhumanist thought. In: *Journal of Evolution and Technology*, 14(1), pp. 1–25.

Cole, T. R. (1992): *The Journey of Life: A Cultural History of Aging in America*. New York: Cambridge University Press.

Collier, T. J. and Coleman, P. D. (1991): Divergence of biological and chronological aging: Evidence from rodent studies. In: *Neurobiology of Aging*, 12, pp. 685–693.

Daffner, K. R. (2010): Promoting successful cognitive aging: A comprehensive review. In: *Journal of Alzheimer's Disease*, 19, pp. 1101–1122.

Dumit, J. (2004): *Picturing Personhood. Brain Scans and Biomedical Identity*. Princeton: Princeton University Press.

Elliot, C. (2003): *Better Than Well: American Medicine Meets the American Dream*. New York, London: W.W. Norton & Company.

European Parliament (2009): Human enhancement study 2009. In: *European Parliament Science and Technology Options Assessment*, http://www.europarl.europa.eu/stoa/publications/studies/stoa2007-13_en.pdf, http://www.humanityplus.org/learn/philosophy/transhumanist-values, date accessed 14 August 2009.

Frazetto, G. and Anker, S. (2009): Neuroculture. In: *Nature Reviews Neuroscience*, 10(11), pp. 815–821.

Giorgi, L. and Luce, J. (eds.) (2007): Special issue: Converging science and technologies: Research trajectories and institutional settings. In: *Innovation. The European Journal of Social Science Research*, 20(4).

Goodwin, J. S. (1991): Geriatric ideology: The myth of the myth of senility. In: *Journal of the American Geriatrics Society*, 39(6), pp. 627–631.

Greely, H., Sahakian, B., Harris, J., Kessler, R.C., Gazzaniga, M., Campbell, P. and Farah, M. J. (2008): Towards responsible use of cognitive-enhancing drugs by the healthy. In: *Nature*, 456, pp. 702–705.

Holstein, M. B. and Minkler, M. (2003): Self, society, and the 'new gerontology'. In: *The Gerontologist*, 43(6), pp. 787–796.

Human Enhancement Study 2009. European Parliament Science and Technology Options Assessment. Available at http://www.europarl.europa.eu/stoa/cms/home/publications/studies?year=2009&q=Human+Enhancement&studies_search=Send+foresp%C3%B8rgsel (accessed 27 May 2014).

Humle, T. and Friislund, M. (2010): 'Study drugs' vinder frem på universiteter, information 1, section 4 [journalist study about the use of so-called 'study drugs' in Danish universities. I have access to the study data through personal contact].

Kampf, A. and Botelho, L. A. (2009): Anti-aging and biomedicine: Critical studies on the pursuit of maintaining, revitalizing and enhancing aging bodies. In: *Medicine Studies*, 1, pp. 187–195.

Kirk, H. (2008): *Med hjernen i behold – Kognition, træning og seniorkompetencer.* Copenhagen: Akademisk Forlag.

Liberati, G., Raffone, A. and Belardinelli, M. O. (2011): Cognitive reserve and its implications for rehabilitation and Alzheimer's disease. In: *Cognitive Processing*, 13(1), pp. 1–12.

Littlefield, M. M. and Johnson, J. (eds.) (2012): *The Neuroscientific Turn: Trandisciplinarity in the Age of the Brain.* Ann Arbor: University of Michigan Press.

Lock, M. (2005): Alzheimer's disease: A tangled concept. In: McKinnon, S. and Silverman, S. (eds.) *Complexities: Beyond Nature & Nurture.* Chicago and London: The University of Chicago Press, pp. 196–222.

Mehlman, M. J. (2004): Cognition-enhancing drugs. In: *The Milbank Quarterly*, 82(3), pp. 483–506.

Petersen, A. R. (1996): Risk and the regulated self: The discourse of health promotion as politics of uncertainty. In: *Journal of Sociology*, 32(1), pp. 44–57.

Petersen, A. R. and Wilkinson, I. (2008): Health, risk and vulnerability: An introduction. In: Petersen, A. R. and Wilkinson, I. (eds.) *Health, Risk and Vulnerability.* New York: Routledge, pp. 1–15.

Reuter-Lorenz, P. A. and Lustig, C. (2005): Brain aging: Reorganizing discoveries about the aging mind. In: *Current Opinion in Neurobiology*, 15, pp. 245–251.

Roco, M. and Bainbridge, W. (eds.) (2002): *Converging Technologies for Improving Human Performance*, http://www.wtec.org/ConvergingTechnologies/Report/NBIC_report.pdf, date accessed 29 May 2009.
Rose, N. (2003): Neurochemical selves. In: *Society*, 41(1), pp. 46–59.
Rose, N. (2007): *The Politics of Life Itself.* Princeton: Princeton University Press.
Rose, N. (2009): Normality and pathology in a biomedical age. In: *Sociological Review*, 57, pp. 66–83.
Rowe, J. W. and Kahn, R. L. (1987): Human aging: Usual and successful. In: *Science*, 237(4811), pp. 143–149.
Rowe, J. W. and Kahn, R. L. (1998): *Successful Aging*. New York: Pantheon Books.
Savulescu, J. and Bostrom, N. (eds.) (2009): *Human Enhancement.* New York: Oxford University Press.
Whitehouse, P. J. and George, D. (2008): *The Myth of Alzheimer's.* New York: St. Martin's Press.

# 5
# The Making and Unmaking of Deaf Children

*Sigrid Bosteels and Stuart Blume*

## Introduction

Most parents-to-be hope for, and expect, a 'normal' baby: perfectly formed, with all its organs, limbs, muscles, and senses, and equipped to follow a normal process of growth and development. Advances in prenatal genetic testing, embryo selection, and assisted reproductive technologies seem to bring the possibility of a 'designer baby' within reach: a child even more precisely tailored to its parents' dreams (Rothschild 2005). This possibility, of course, is one that does not find universal acclaim (e.g., Parens and Asch 1999). Clearly, for most parents-to-be the child they hope for will be able to see and to hear. Since the capacity to hear develops midway through a pregnancy, it is not uncommon these days for women to encourage sound perception by wearing tinkling bells or playing music to their unborn child.

Because hearing is taken to be the basis of language acquisition, it is considered important to establish the presence of normal hearing in a baby. In the richer countries of the world, including Western European welfare states such as Belgium and the Netherlands, all babies' hearing has long been tested as a matter of routine, though the way in which testing is carried out, and when, has changed over the last few years. Where hearing is found to be deficient by reference to population norms, and the child categorized as 'deaf' or 'hard of hearing', some kind of prosthesis will almost certainly be prescribed. Here too, technological advances have led to new prosthetic options. Medicine, with its techniques of measurement, its categorizations and prosthetic devices, intervenes profoundly – and perhaps increasingly so – in the lives of deaf children and their families.

The medical profession, in particular the ear, nose, and throat (ENT) doctors and the audiologists who carry out these procedures, would not recognize their work as 'making' or 'unmaking' deaf children. Rather, they would say that with the aid of increasingly sophisticated technologies, they 'diagnose', 'assess', and 'compensate' for hearing loss. What is the difference between their terminology and that used here? By using the term 'making' in place of 'diagnosing' deafness in children, we want to emphasize the complexity of this social process that includes a restructuring of the work of parenting and new sets of interdependencies, which the diagnostic act serves only to initiate (Charmaz 1991). We speak of 'making' deaf children because, as Scott showed long ago for blind people (Scott 1969/1991), this is only the first step in a much longer process in which the deaf/blind child's family learns its responsibilities and acquires the necessary skills, a variety of professionals and institutions are mobilized in support, and the child learns to behave as a deaf/blind child is expected to.

Medical professionals would perhaps be still more uncomfortable with the notion of 'unmaking'. It suggests something far beyond what medical science now has to offer. 'Of course we can't *cure* deafness', many would indignantly retort, 'we can only alleviate some of its effects'. 'Unmaking deafness' is not the language of the medical profession, though it is the language of the mass media, which avidly and often overenthusiastically magnifies the more modest claims that circulate in professional worlds (Dresser 2001). However, although 'unmaking' can be taken to be synonymous with 'curing', our use of it here is intended to indicate a wider set of associations, as we will critically investigate by including 'parents' voices.

## Making

As the many guides to 'successful parenting' imply, new skills have to be acquired, new sensitivities developed, attention refocused. Parents and their babies discover the new sensations of family life and become embedded in an intimate but also social world of interaction. Mother or father and daughter or son embark on a lifelong socialization process, implying re-socializing experiences for parents as their children grow up. One becomes a parent in the process of educating and guiding one's children and, in so doing, new rights, responsibilities, obligations, and dilemmas accumulate. All of these have a particular salience and qualities in the case of parents whose child is diagnosed with hearing loss.

A phenomenological view of the lived experiences of parents with children with a disability challenges us to look at the everyday world from different angles and to interrogate the multiple meanings of early interventions. Parents with deaf children are confronted with what Landsman (2003) and Larson (1998) describe as a paradox in negotiating parenthood. Opposing forces are at play between saying to the child 'I love you as you are', while also expressing 'I would do anything to change you' (Landsman 2003, p. 1949). In case of childhood deafness, the infant's lack of auditory impulses is considered to be a direct cause of possible parental problems and induces early dependency on technical and medical support. From a professional and policy point of view, the possibility to hear and speak is categorically set as the standard for a normal, happy, and healthy child. Accordingly, the absence of these capacities creates a generalized identity construction of the child as deaf or disabled, with the implied risk of diminished personhood.

How a child's defective hearing is established has changed over time. In the latter part of the 20th century the initial test, generally performed at around 9–12 months, involved children being placed on their mother's lap in an acoustically adapted room. Two professionals were present, one in front of the child and one positioned behind. The person in front kept the baby visually engaged by showing it all sorts of toys (mostly building blocks), while the one at the back introduced auditory signals such as the sound of a rattle, a spoon in a coffee cup, a tambourine, and so on, alternating these sounds to the left and right. Did the child react? This test led to relatively high percentages of false positives when the child failed to react as expected, which caused unnecessary tension for parents. Further tests would be done, the child's hearing would be measured, and an audiogram (a representation of the intensity of a just-audible sound at each of a range of frequencies) produced. If there was a significant deviation from 'normal', the child was diagnosed as hard of hearing or deaf. Other tests might also be done to determine the nature and cause of the hearing loss.

A new form of hearing test has been introduced in the last decade or so. Instead of being tested at nine or ten months, a baby's hearing can now be tested when it is just a few weeks old, using a sophisticated technique based on so-called auditory brainstem response audiometry. The Flemish region of Belgium began to use what is generally known as the 'Algo test' routinely in 1998, before any other European country or region. Why? The national public childcare institution *Kind & Gezin* ('Child and Family') – responsible for detecting hearing loss in young children since the late 1970s – gave the motivation for their pioneering efforts in this early healthcare practice as follows:

> Hearing impaired children miss sensory stimulation, which is a prerequisite for speech development. This impairment has a fatal influence on the total development of the personality in all its social, emotional, intellectual and locomotive aspects. Furthermore, the absence of auditory impulses has a negative effect on the process of education and on the parent–child relationship.
>
> (Van Kerschaver and Stappaerts 2008, p. 4. Our translation)

Policymakers and the head of the medical department of *Kind & Gezin* were convinced that identifying hearing loss in children as early as possible has therapeutic and pedagogical value, and is also cost-effective. Once hearing loss has been diagnosed, parents can be instructed on the responsibilities that bringing up a deaf child entails, and encouraged to discharge these responsibilities diligently. Parenting is in effect redefined on the basis of techno-medical rather than affective priorities. In particular, because of the importance of adequate language and speech acquisition, social-emotional development, and future educational opportunities for young children, the justification for technical and biomedical interventions is formulated as self-evident, with speed being of the essence.

## The techno-medical trajectory: Towards therapeutic parenting

In Belgium, hearing screening is carried out by nurses when babies are between four and six weeks old. Every parent of a newborn baby is visited by a district nurse from *Kind & Gezin*, with the only expectation being that the baby should be calm or asleep if possible. The two electrodes of the portable Algo test device are placed on both sides of the baby's head and they transmit signals to the brain. After a few minutes the results are displayed: either 'pass' or 'refer'. A 'pass' result means that normal hearing levels have been found. A 'refer' result means that hearing is possibly impaired, and further assessment is necessary. A second test, in the presence of a medical doctor, is typically performed no more than 48 hours after the first one. If still positive, parent and child are referred to one of the specialized hearing centers (often the ENT department of a hospital) where additional diagnostic procedures and follow-up are planned.

## How do parents experience being confronted with the diagnosis of their child's deafness?

To illustrate these parental experiences we draw on a qualitative study based on retrospective interviews with 16 parents of congenitally deaf children carried out by Bosteels in Flanders (Belgium) between 2005 and 2006. All names used in the quotations are pseudonyms.

Parents for whom an initial 'refer' result brought anxiety and the prospect of further medical encounters were often upset about how the test was carried out and the results communicated.[1]

> That was the first time the nurse experienced this because the Algo test hadn't been around for very long, just a few months or less than a year. She says 'there must be something wrong with the appliance [...]. I will visit you at home on Thursday and bring another one.' Then here, with that other machine, again 'refer'. She then spoke on the phone to someone from the company of the Algo test, did the test three times and by then I started to get worried. What's going on here?
>
> (Mother of Jolien, Belgium)

The insistence on parents deciding and acting quickly in response to a 'refer' result from the Algo test is sometimes too much to handle. Less and less time and space are left for dealing with feelings of doubt, insecurity, anger, denial, and hope. Under the influence of this medico-technical trajectory, a tension emerges in the parenting process. The success of newborn hearing screening and the subsequent follow-up programs stands or falls with the confidence and trust of parents. If they misunderstand or neglect professional messages, delays in or denial of subsequent care may follow. From the moment of diagnosis, all the attention is focused on the functioning or failure of the capacity to hear, and parents are overloaded with new information and new obligations. Parents may forget the entire, healthy baby with a non-life threatening condition and concentrate on what is missing or threatening.

> You do all kinds of tests yourself, all the time. She would be asleep in her cot and you make noise in all kinds of ways to see if she jumps... Like with two pan lids; she did not jump as such but she did blink. But that was because of the draft you are making... At that moment you think, she has heard it already... You fool yourself in so many ways, you cannot accept it.
>
> (Mother of Jolien, Belgium)

The identity of the child becomes fixed by its categorization as a deaf or hard of hearing child, and parents are supposed to nurture and educate their child accordingly. The happiness at becoming a parent may soon turn into feelings of loss and sadness at becoming the parent of a deaf child. Because few parents have any prior experience of deafness, recommended ways of interacting and communicating are experienced as unfamiliar and unnatural.

> You are at home, you just had a baby and then...your child can't hear. You say its name and you realize, 'Oh, he cannot hear me.' That's terrible. It may be stupid but those first three weeks I was just going on at the child while he couldn't hear me at all.
>
> (Mother of Bram, Belgium)

The professional information they receive as they embark on this medico-technically shaped parenting trajectory offers parents the hope of their child's entry into the world of oral communication. The child should be exposed to sound as much as possible. The use of hearing aids is the first step to be taken, regardless of the degree of hearing loss. Even children with a profound hearing loss (who may ultimately receive little benefit) are supposed to wear them right from the start. Fitting these technical aids to their baby isn't what parents had dreamed of.

> Well a child, euh a baby with hearing aids, is that really necessary? [...] The whole family was startled and they said that it can't be, that he was reacting all right and it will turn out ok, you'll see. They said, 'You will not do it, will you, a baby with hearing aids?' But I said, 'I have to, it should.' [...] People at the hearing center over there went on and on about it and convinced us two hundred percent, because that child must talk. And if you wait for two or three years he won't be able to speak, it would only be babbling. Indeed, if you pay attention to deaf people that were born before and who weren't looked after properly, they stay deaf.
>
> (Mother of Wouter, Belgium)

However hesitant their response to the technical interventions required may be, all parents are nevertheless confronted with the intensity and velocity of the medico-technical roller coaster. Sometimes they begin to approach their nurturing tasks more as therapists than as the kind of parents they had probably hoped to be.

> Jolien had hearing aids when she was six weeks old. She is the youngest ever, they know all over the world...If I think back on it

> all now, but you have no idea. And that was that, at first every other week at least, every ten days, for new ear molds. Because those little ears grow very quickly. And as soon as there was any air between the ear molds, they began to whistle. So you couldn't cuddle your child, or pick her up properly because that hearing aid would always be whistling. We were told, 'Spend as much time as possible, expose her to noise,' so in our free time, yes, it became an obsession, we did nothing else... But you end up being a therapist, you're no longer a mother or father.
>
> (Mother of Jolien, Belgium)

This 'therapeutic parenting' imposes new constraints and new obligations as parents and child learn how to be a 'deaf family'. As they do so, intimate family spaces become filled with expensive hearing technology and with well-meaning but unfamiliar professionals. Through continuing emphasis on the regular practice of teaching and rehabilitating a child considered in need of repair, the child's affective needs may be neglected.

Families are often confused and insecure, not only about the adjustments that are needed for their individual child but also about the responses of the outside world. Through the gaze of others, a son or daughter may become a source of shame.

> They said, 'Your child is going to be helped.' You should be grateful that you are going to be helped, I know. And, about a week before, we saw a documentary about cochlear implants and I said, 'Ok, if this is what is going to help her, I'll put her into the hospital, stick such a thing underneath her hair and you won't see it.' Yes, but no, it doesn't work for her. It had to be hearing aids. I thought, my goodness no, such a little child, such a little face, to such little ears they will attach those bastards of hearing aids. Also because of what other people might say about that. The first moments, I thought it was horrendous and they looked gigantic, you saw them coming out underneath her hair. If it could be, she wouldn't wear any. Of course there is no alternative. But if there comes a day when they say it is possible to operate on her so that she could hear like any other child, then we'll throw those hearing aids out, immediately.
>
> (Mother of Sien, Belgium)

Occasional counter-narratives show seeds of agency, of resistance to a professional discourse guided by an instrumental logic that challenges parents to act and not reflect. A few parents with the resources, skills, and self-confidence to reflexively seek their own solutions rather than unquestioningly follow medical advice tell a different story. They too are engaged in 'making deaf children', but they do so differently. Kobe and Marthe's mother, who once trained in speech therapy and audiology (though she never practiced professionally), explained her situation as follows:

> Everybody always had to laugh when they came into the house. There were papers and photos everywhere. There was a great big mirror that I'd bought at the flea market. I was always sitting on the ground in front of that mirror, playing with them, fetching things…or going with them to the shop, so that they could feel, taste, and hold things. I wasn't being a therapist, you know, it was all playful…Something that really made a difference for us was that we started to follow sign language classes. We did that for four years. The children's grandparents too. That was really important for the whole family, in terms of acceptance, of learning to cope with a handicap. Just to learn what it means to be deaf. To learn to accept things, and not to be scared. I mean…if you're not scared by your child's handicap you'll be able to accept it much better.
>
> (Mother of Kobe and Marthe, Belgium)

Although this mother too is engaged in a work of construction, in which the whole family participates, her account displays an agency that is less apparent in most of the other parents' stories. On the one hand, the family has chosen to learn sign language, a course of action that the medico-therapeutic trajectory tends to exclude – either by forbidding or discouraging it. On the other hand, the manner of telling does not suggest that the affective dimension of parenting has been sacrificed to the dictates of a therapeutic regime imposed from outside. Relatively few parents have the skills and the self-confidence to escape a regime that requires diligent conformity rather than considered reflection, and which seems to make the affective dimension of parenting a lesser priority. These parents too are helping to make deaf children, but perhaps in a different way and, perhaps, with different goals.

## Unmaking

Most hearing parents of children diagnosed as having a hearing impairment allow themselves to be led onto a medico-technical trajectory, a trajectory of therapeutic parenting. Why? Medical specialists and audiologists do not generally claim that deafness can be 'unmade', and parents are not encouraged to think that their deaf child can be made into a hearing child. The more modest indication is that, with sufficient dedication and application, the child will learn to use what little residual hearing it has to optimum benefit and so will learn to function in the hearing world.

In recent decades, the social and medico-technical context within which the parenting of deaf children takes place has become more complex and contested. This is partly to do with the effects of earlier diagnosis, referred to above. The therapeutic trajectory has been reshaped by developments in screening technology, leaving less and less time for the affective aspects of child-rearing. It is also partly to do with developments in prosthetic technology. In the course of the 1990s a new device, the cochlear implant, became widely available.[2] Implantation, at an age that has declined from years to months, is becoming the widely recommended next step in therapeutic parenting. Just as a medical act (diagnosis) initiates a transformation of parenting, so too a surgical act (implantation) initiates a further set of tasks and obligations.

Although medical professionals may avoid making claims about being able to unmake deafness, the public discourse that accompanied the introduction of the cochlear implant into medical practice did involve such claims, either explicitly or implicitly. In the mass media the device soon became known as the 'bionic ear', a term that evokes a quite different image than 'cochlear implant'. For one thing, it seems to invoke the limitless potential of science. This was not accidental. The term 'bionic ear' was coined in Australia with precisely these associations in mind, in the course of planning a televised fundraising appeal for the Melbourne implant program. Someone had apparently suggested that it was as though they were building an ear for the 'Six Million Dollar Man', the eponymous character from a popular television program at the time (Clark 2000, p. 71). Though there were some doubts about a term that seemed to hint less at corrected hearing than at enhanced hearing, it proved irresistible. It appealed both to implant surgeons looking to mobilize resources and to the mass media looking to reflect popular faith in medical science. Melbourne soon became home to the 'Bionic Ear Institute'.[3]

The media discourse surrounding a new medical technology is important. People's expectations of medical technologies are not formed only, and perhaps not even principally, by the information provided by medical professionals. Growing up in a culture that expects so much of medical technology, most North American and European parents will have spent their lives exposed to the consistent message of hope reiterated over and again by the mass media. As far as the cochlear implant is concerned, that message, often supported by personal accounts of successful implantation, has been a clear and simple one: the cochlear implant can turn a deaf child into a hearing one. From his professional experience in conducting pre-implantation psychological assessments, Robert Pollard came to the conclusion that candidates and their families often 'present with oversimplified or distorted perceptions about cochlear implants based on material they have been exposed to in the popular media' (Pollard 1996, p. 22). Furthermore, while medical professionals could play an important role in providing comprehensive information, helping parents to make a reasoned and informed choice, that is rarely what they do. Looking at the information materials available to medical professionals, and on which they base their advice, Pollard finds references to deafness resulting in 'social isolation', 'depression', and 'personal danger' (Ibid., p. 22). He continues, 'Knowing little else about the ordinary lives of deaf people, what parents would leave such a presentation with anything other than the message that, if they love their deaf child, they should pursue cochlear implantation in order to prevent such catastrophes.'

However much they try to resist succumbing to media hype, parents may still have difficulty avoiding the simple picture painted by the mass media, as refracted through the views of others close to them.

> One or two friends have said, you know, 'Has he not been able to have a cochlear implant?' They've worded it in such a way as if we've been denied something. So I think the press do... it is portrayed as 'This is a cure for deafness'.
>
> (Mother of Peter, England)[4]

The evocations of deafness unmade, of bionic hearing, that characterized early media reporting of cochlear implantation were contested. In a number of countries organizations of parents of deaf children were unhappy with it (Blume 2010, pp. 69–71). Since, in their view, it was far from clear to what extent deaf children could profit from the implant, it was necessary to proceed with caution. They were worried that publicity around the 'bionic ear' would evoke unrealistic hopes and expectations. Their concerns, in the early years of pediatric implantation, focused principally on uncertainty, on the lack of knowledge.

Later critique took a different turn. Implant programs in many countries insisted (as many still insist) that parents commit themselves to the exclusive use of spoken language. Through emphasis on the use of speech at home, and attendance at an oral school, the value of the implant is maximized, and the child is enabled to participate in mainstream hearing society. Critics have argued that such a practice is not in fact in the child's best interest: that other, social and psychological, considerations are thus neglected, and that use of sign language can be a preferable alternative (see Crouch 1997).

In the last few years a number of empirical studies of parents' expectations regarding the cochlear implant, and their experience of the decision-making process, have been published. The picture that emerges from most of these studies is a surprisingly bland one, little marked by doubt or controversy. Mostly, and perhaps unsurprisingly, they conclude that in opting for a cochlear implant parents were largely motivated by the desire to have their child function as a hearing person. Typically, parents thought themselves to have been well-informed, but very rarely did this involve trying to understand how deaf people live their lives or the nature of the objection to the implantation of deaf children. For example, a British study showed that the majority of British parents had found the decision for implantation relatively straightforward, as they believed their child had nothing to lose and everything to gain. There was 'really no alternative'. Most parents

> believed themselves to have been well-informed about the intervention, and that there was nothing of relevance which they had not considered before sanctioning implantation. 'The information we've had has been first class all along the way. We felt we were in very safe hands.'
>
> (Sach and Whynes 2005, p. 405)

Many parents of deaf children in the various countries in which these studies were conducted were well aware that there are different views regarding pediatric implantation. But consider how the decision-making process works for many families. The doctors and other professionals on whom the parents depend tell them about the cochlear implant, somewhat along the following lines:

> There is no certainty that it will make him hearing, but it will help. It will make him hard of hearing. We know because thousands of children worldwide have been implanted and we know how much they've benefited. It will help him enormously in acquiring (spoken) language, although we can't predict exactly how much it will help

> him. To get the most out of the implant it is better to avoid using sign language (though that is of course a perfectly respectable means of communication for those who cannot speak or hear).

The parents contact the cochlear implant center, and are provided with all kinds of information regarding the implant. The information provided also explains that deaf people do not see themselves as in need of repair. The parents are told that, from the deaf point of view, sign language is an acceptable and alternative means of communication to which their child has ready access. Many of the parents feel they now know enough. They feel no need to look further into what deaf people have to say, or the possibility of starting to learn sign language. Their child is entered into the implant center's selection procedure. The professionals they have met clearly have the children's interests at heart, and trusting them comes naturally.

A few parents, however, proceed more cautiously. The father of Jelle, a Belgian child, decided to learn sign language when his deaf son was a few months old. It proved difficult to find opportunities for practicing, but his decision to stop learning was provoked by something else. Jelle's mother explained it as follows:

> My husband went to follow [sign language classes] in H [name of town], in a room attached to a café. It was with real deaf people. But one evening he was sitting with two deaf people, and they were really against the idea that we were going to let our Jelle be fitted with a cochlear implant. They said something like 'You've got to adjust to us, not us to society'. They were really against it. It was really extreme, that café. There was a notice on the door: 'Only for deaf and hard of hearing'.
>
> (Mother of Jelle, Belgium)

A few parents, like Jelle's father, do overcome their fears and go along to the local deaf club. Nobody speaks much there, they cannot follow the signing, and it is not easy to make contact. They learn that it is possible to enroll for a course in sign language, and a few do. Persevering with their sign language lessons, they get to know a few deaf people. Gradually, they feel they come to understand what the life of a deaf person is like, and what lies behind the objections of culturally deaf people to pediatric implantation.[5] They realize how little most people (the doctors they have spoken to, teachers, their friends and neighbors) know about deaf people or the Deaf community. Culturally deaf people tend to be rather scornful of doctors' limited understanding of deafness. But the more the parents come to understand, the more difficult it is to be sure of what to do. Some of the parents who have made an effort to approach the local deaf association feel inhibited (by their own inability to sign), or even feel that they have been rebuffed. Yet others go further – learning sign language, developing contacts with the Deaf community – but nevertheless, despite knowing deaf people's objections to implantation, still consider cochlear implantation for their child. Here there is no question of rejecting deafness, or of being put off by the difficulties of making contact with the Deaf community. As the environment becomes more complex, and as the value of sign language becomes apparent, the choice becomes all the more difficult.

Many or most parents prefer to avoid this difficulty, avoid contemplating the alternative offered by the Deaf community. One of the English parents, the father of Paul, put it like this:

> Myself and [wife], neither of us have been very enthusiastic about sending Paul along to these meetings [for deaf children]. And I think it's because we don't want him to identify himself as a deaf child. We want him to have this sort of consciousness that he is a child, a normal child, who happens to be deaf. And so if we group him together with other deaf children – I know it sounds as though they've got the plague or something, which obviously we know they haven't – we just thought it might sort of reinforce in him that this is a very, that he's a very definite type of child, do you know what I mean? We thought it might stereotype him in a way... It may be partly because neither of us wants to subconsciously accept that he's deaf, I don't know.
>
> (Father of Paul, England)

Other parents, whether they decide on an implant or not, see things differently. What Charlie's mother had to say points to the enormous gap between the simple abstraction underlying what they had been told and their own experience:

> We have for four and a half years – and this is no disrespect to any parties that have been involved – but it's always been a case of, he has got some residual hearing and he will learn to talk, you just need to keep plowing the information in and it's going to come back. So we had spent four and a half years talking, showing, doing, and the frustrations, the older he gets were horrendous. Simple things like drink.

> You know, he'd come in and you'd have to get every cup out of the cupboard to find out which one he wanted, go through every drink. And then we might not have what he wants, but he can't explain what he wants. So, we have another tantrum with throwing things and screaming. And then it suddenly dawned on me, when he was one and a half years old – he needs something more. And if that channel is signing, then that channel has to be signing. Whilst I've been opposed to it thus far, only through again, the information that has been put my way: 'It's not a signing world out there' and blah-de-blah. I suddenly realized that that's been the answer. And it's made our lives so much easier. [...] And I feel angry, for want of a better word, that I haven't done it so much earlier. And I feel that I've wasted a lot of time. And I feel that I've wasted... I feel guilty – because Charlie deserved something before now. So, yes, we are learning to sign.
>
> (Mother of Charlie, England)

## Identity and belonging

Most studies of parents' expectations and experiences, conducted under the auspices of an implantation center, include only parents whose children have previously been implanted there. Neither the parents interviewed, nor the researchers doing the interviewing, appear to have given much thought to the question of the child's identity. An exception is a Japanese study that also included parents who had rejected implants for their children (Okubo et al. 2008). Here too, the most frequently mentioned expected benefit was improved auditory skills, with some expecting that their children's speech would also improve. But in this study, unlike the others, some parents 'expressed concerns that [...] incomplete improvement in hearing and speech would leave children without a clear identity in either the hearing or deaf communities' (Okubo et al. 2008, p. 2441). These parents were worried that an implanted child would be neither hearing nor non-hearing, 'a vague kind of way to exist' (Ibid., p. 2441). Some Japanese parents 'were puzzled by this separation between hearing and Deaf communities, and wondered whether children with implants would form a Deaf identity, hearing identity, or a separate new identity' (Ibid., p. 2442). One of the present authors interviewing Belgian (Flemish) parents a few years after diagnosis and prosthetization (with either an implant or hearing aids), asked how parents then thought about their child. Many found it difficult to find the right words, but their responses were very varied, including: 'a real deaf person', 'a hearing child', 'a hard of hearing kid', and 'a deaf child with a handicap'. But few published studies refer to puzzlements or doubts of this kind.

A small minority of parents seem to have tried, or were able, to make thoughtful and informed decisions, in the sense of truly imagining the available alternatives. Keith's parents, for example, were trying to imagine a future in which their child would be able to move between two worlds:

> It's more in Keith's interest. Ours as well, so that we can understand deafness. It's all well and good having a deaf child, but you've got to understand the way deaf people think and feel. You can't just sort of ignore it, because it's always going to be there. And the more we know about it, the better. I mean these are the people who have had the experience all their lives and they could tell you the problems you're going to come against. The pitfalls, how they feel, if they get depressed or whatever.
>
> (Mother of Keith, England)

That few parents attempt to imagine alternatives in this way has partly to do with the very different sources of information available to them. Detailed information on the cochlear implant and what it might offer their child is provided by the implant teams: enshrined in a web of institutional practices and carefully cultivated relationships of trust. These parents are accustomed to trust the medical and other professionals on whom they are so dependent. 'They are there to do their best for us', many might say. When, as in the case of Charlie's mother, quoted earlier, professionals appear to have betrayed their trust, the sense of betrayal is powerful and painful. By contrast, what deaf people have to say, for those who try to find out, bears the stigma of its origins at the margins of society. There is also a difference in cultural dispositions that shapes parents' receptivity to the conflicting messages. The mass media, emphasizing the promise of medical progress, reinforces the hope of restoring normality: a normal (meaning hearing) child, a normal family. Media representations rarely encourage an imaginative contemplation of alternative ways of 'being in the world', or of 'moving between two worlds'.

The reason that the medical perspective on deafness influences parents so much more powerfully than that of deaf people themselves reflects more than status differences in the sources of the two messages. It also has to do with the challenging and unfamiliar nature of the argument underlying what the Deaf community has to say. Part of that argument is that it is in the individual child's best interest to grow up using sign language, as a member of the Deaf community (Crouch 1997). Some, however, go further. Psychologist/historian Harlan Lane and professor of Deaf Studies Benjamin Bahan have argued that a deaf child 'in the normal course of things' would become a member of the Deaf community and so have values different from his or her parents. Children born deaf are said to have 'a Deaf heritage from birth' as a consequence of their physical constitution (Lane and Bahan 1998). Their cultural status is determined by the culture the child would enter 'in the normal course of things'. Noel Cohen, a well-known New York cochlear implant surgeon, disagrees with such a stance: 'Deaf children of hearing parents are not members of the deaf community until they are either placed in that community by their parents or voluntarily decide to enter it' (Cohen 1994, p. 1). For Cohen and his colleagues – and for many parents – implantation is seen as offering the child the possibility of later choosing whether or not to enter the Deaf community. As far as Lane and Bahan are concerned, what is at stake is whether or not the child is allowed to take possession of his or her birthright. In other words, the crucial issue here, according to these critics, is fundamentally a matter of identity. From this perspective, 'unmaking deafness' is not a matter of cure but of denial: the denial to the child of its cultural heritage.

These arguments – the notion of a 'Deaf community' and a 'Deaf identity' as a child's natural heritage – can be threatening and disruptive for parents. That adult deaf people might have useful knowledge is one thing, but the notion of a strange community, distant, different, difficult to approach, claiming some sort of affinity with the child, is something else entirely. Introducing notions of complexity and fluid identities into family life, they challenge taken-for-granted ideas about the integrity of the family, parent–child relationships, and the agency of the child.

## Conclusions

What does medicine offer the parents of children diagnosed as deaf or hard of hearing, as well as the children themselves? And what does it deny them?

The screening of newborn children for a growing range of (largely genetic) conditions is becoming more and more common. While the justification is that families will welcome information about such conditions, 'little is known about how those technologies actually affect families' lives' (Timmermans and Buchbinder 2010, p. 408). Knowledge of how a new diagnostic modality might affect the lives of those tested, their families, or the quality of parenting has not been a matter of primary concern. Introduction of new diagnostic or screening tests has been dependent largely on medical perceptions of the seriousness of the condition to be tested for and, perhaps, on the possibility of intervention. Where necessary, additional justifications can be sought in 'consumer demand' or 'cost effectiveness'.

But unlike the vaguely defined genetic conditions that concern Timmermans and Buchbinder, a diagnosis of hearing impairment inevitably leads to unambiguous professional follow-up. When routine screening indicates the possibility of a hearing impairment, parents are encouraged to enter a trajectory of 'therapeutic parenting'. Knowing nothing of deafness, scared of what it might mean for their child and for the family as a whole, bewildered, and guilty, most are relieved to do so. The medical professionals they encounter clearly have their child's interest at heart, and trust in the medical profession comes naturally. Gradually, parents become familiar with the responsibilities, and the dilemmas, that parenting a hearing impaired child entails. They come to rely on the assistive devices provided, and on the support offered by professional caregivers. In this way hopes are sustained: the hope that the child will be able to function normally in society, and that some of the shame attached to parenting a child that is different can be avoided. This trajectory, however, is changing.

Through a variety of mechanisms, the objectives, assumptions, and commitments of medical practice exercise a profound (perhaps unequalled) influence on science and on the development of new technology. One of these commitments, in the field of hearing, is that the earlier the intervention, the greater the chance that the effects of defective hearing can be compensated. Facilitated by new technology, the trajectory on which parents enter has become a veritable roller coaster. We have focused on two such changes, brought about by the introduction of the Algo hearing test and the cochlear implant. Responding and intervening as rapidly as possible is emphasized, and parents are warned of the consequences of failure to do so. There is less and less time for the affective aspects of parenting, and (as Jolien's mother put it) 'you end up being a therapist, you're no longer a mother or father'. There is also less and less time for reflection. Trying to create time and space for critical reflection requires a degree of resistance for which few parents seem to have the desire or the confidence. In the course of the last two decades the cochlear implant has become a new therapeutic option. Here too, rapidity of intervention is emphasized, and parents are urged to decide quickly. The recommended age of implantation has been reduced from years to months. Parents' hope that, thanks to the implant, their child will be able to function as a 'normal hearing person'.

The accounts of a few parents point to what resistance and critical reflection might entail: a wish to avoid such a therapeutic experience of parenting and, most significantly of all, consideration of the possibility of learning sign language. A few parents in our studies thought that they needed to know more; in particular about what sign language might have to offer them and their children – though this is typically discouraged by many implantation teams. The principal difference between the majority of parents, those who unhesitatingly and uncritically embark on the medico-therapeutic trajectory, and the minority, who create space for consideration of alternatives, consists in contrasting senses of how the quality of the child's later life is understood. For the majority, inflected by what medical practice is able to provide, it is a matter of hearing and speaking. A good life depends upon participating in the oral communication of the world at large. The minority are less certain. These parents wonder about the child's future identity, its sense of belonging, the possible benefits of being able to function in two worlds. Some, at least, of the arguments of Deaf advocates make sense to them: arguments to the effect that a good life requires a strong sense of self and of belonging, and that it is through membership of the Deaf community, rather than functioning at the margins of the hearing world, that these psychosocial benefits are acquired.

As we have emphasized repeatedly throughout this paper, few parents in Western industrial societies seem willing and able actively to contemplate these alternatives. We have suggested that the reasons for this have partly to do with the sources of the conflicting messages, and partly to do with the notion, discomforting to most hearing parents of deaf children, that their deaf child in some sense 'belongs' to a community of which they themselves are not – and cannot be – part. How could it be different?

Writing about disability, anthropologists Rayna Rapp and Faye Ginsburg (2001) suggest that what is needed is something they call 'rewriting kinship'. They draw a distinction between the public discourse surrounding disability on the one hand, and 'the daily and intimate practices of embracing or rejecting kinship with disabled fetuses, newborns, and young children' on the other (Rapp and Ginsburg 2001, p. 533–4). The first is a matter of ethical debates about reproductive choice, of legislating for access to buildings, schools, employment. In other words, it is comparable with the political claims of the Deaf community, though this is something with which many hearing parents have great difficulty. The second, rooted in changing daily practices in the family, in the work of clarifying and articulating what a disabled, or deaf, child might mean for the family as a whole, is more fundamental. At the heart of the notion of citizenship championed by the disability rights movement is the integration of disability into everyday life. The shift from exclusion to inclusion, reshaping the possibilities of life as a deaf person (or person with a disability) has to have its roots in family life.

Rapp and Ginsburg argue that 'public storytelling', testifying to problems overcome, lives successfully led, is crucial to bringing this integration about. This public storytelling must itself be embedded in a broader debate. As philosopher Robert Sparrow argues, in relation to prenatal genetic testing for deafness, "developing good policy [...] will require thinking hard about what sorts of experiences and achievements make a human life worthwhile and about the relationship between our ideas of what is normal and the availability of these goods in a world in which we have the power to shape the capacities of those we bring into the world" (Sparrow 2010, p. 464).

## Notes

1. Because of the relatively low prevalence of congenital deafness (1.4/1000), nurses were not regularly confronted with positive test scores and had to become acquainted with these new tasks.
2. The cochlear implant consists of an externally worn microphone and signal-processing device, and an electrode that is surgically implanted into the inner ear, or cochlea. The device enables auditory stimuli to be transmitted to the brain, even in people whose deafness is such that they do not profit from the amplification provided by conventional hearing aids.
3. Now the Bionics Institute.
4. Interviews with parents in the south of England were carried out in 1996–1997 in collaboration with, and under the supervision of, Professor Lucy Yardley (now of Southampton University).
5. Many authors distinguish between culturally deaf people who use a sign language (capitalized as Deaf) and those who are audiologically deaf (uncapitalized). Here we use the capitalized form only when referring to the Deaf community.

## References

Blume, S. (2010): *The Artificial Ear: Cochlear Implants and the Culture of Deafness.* New Brunswick: Rutgers University Press.

Charmaz, K. (1991): *Good Days, Bad Days. The Self in Chronic Illness and Time.* New Brunswick: Rutgers University Press.

Clark, G. (2000): *Sounds From Silence: Graeme Clark and the Bionic Ear Story.* St Leonards, NSW: Allen and Unwin.

Cohen, N. L. (1994): The ethics of cochlear implants in young children. In: *American Journal of Otology*, 15, pp. 1–2.
Crouch, R. A. (1997): Letting the deaf be deaf. Reconsidering the use of cochlear implants in prelingually deaf children. In: *Hastings Center Report*, 27(4), pp. 14–21.
Dresser, R. (2001): *When Science Offers Salvation. Patient Advocacy and Research Ethics*. Oxford: Oxford University Press.
Landsman, G. (2003): Emplotting children's lives: Developmental delay vs. disability. In: *Social Science & Medicine*, 56, pp. 1947–1960.
Lane, H. and Bahan, B. (1998): Ethics of cochlear implantation in young children. A review and reply from a Deaf-world perspective. In: *Otolaryngology Head and Neck Surgery*, 119(4), pp. 297–313.
Larson, E. (1998): Reframing the meaning of disability to families: The embrace of a paradox. In: *Social Science & Medicine*, 47(7), pp. 865–875.
Okubo, S., Takahashi, M. and Kai, I. (2008): How Japanese parents of deaf children arrive at decisions regarding cochlear implantation surgery. A qualitative study. In: *Social Science & Medicine*, 66, pp. 2436–2447.
Parens, E. and Asch, A. (1999): The disability rights critique of prenatal testing: Reflections and recommendations. Special Supplement, In: *Hastings Center Report*, 29(5), pp. 1–22.
Pollard, R. Q. Jr. (1996): Conceptualizing and conducting preoperative psychological assessments of cochlear implant candidates. In: *Journal of Deaf Studies and Deaf Education*, 1, pp. 16–28.
Rapp, R. and Faye Ginsburg, F. (2001): Enabling disability: Rewriting kinship, reimagining citizenship. In: *Public Culture*, 13(3), pp. 533–556.
Rothschild, J. (2005): *The Dream of the Perfect Child*. Bloomington: Indiana University Press.
Sach, T. H. and Whynes, D. K. (2005): Paediatric cochlear implantation: The views of parents. In: *International Journal of Audiology*, 44, pp. 400–407.
Scott, R. A. (1969/1991): *The Making of Blind Men. A Study of Adult Socialization*. New York: 1969, Russell Sage Foundation; 1991, Transaction Books.
Sparrow, R. (2010): Implants and ethnocide: Learning from the cochlear implant controversy. In: *Disability & Society*, 25, pp. 455–466.
Timmermans, S. and Buchbinder, M. (2010): Patients-in-waiting: Living between sickness and health in the genomics era. In: *Journal of Health and Social Behavior*, 51, pp. 408–423.
Van Kerschaver, E. and Stappaerts, L. (2008): *Jaarrapport Gehoor 2008. Universele gehoorscreening in Vlaanderen. Doelgroep, testresultaten en resultaten van de verwijzingen* [Annual report on universal hearing screening in Flanders]. Brussels: Kind & Gezin.

# 6

# Token of Loss: Enthography of Cancer Rehabilitation and Restoration of Affected Lives in Kenya

*Benson A. Mulemi*

Both cancer, which can be life-threatening, and its treatment, which can be life-saving, contribute to bodily degeneration and social and emotional disruption. Rehabilitation through biotechnology can support new lives and assist people in coping with any acquired disability in the process of and after treatment. This entails repair, which utilizes different sorts of objects to improve the body's natural capabilities in its new state. Cancer patients' experiences of rehabilitation may vary from the definition of rehabilitation offered by professional and family healthcare givers. Rehabilitation of cancer victims generally involves helping them to obtain maximum physical, social, psychological, and vocational functioning and independence within the limits imposed by the disease and its treatment (DeLisa 2001, p. 770; Romano et al. 2006). The techniques for restoring well-being may even endow the persons using them with extra abilities that they may not have had before diagnosis of the disease, treatment, and rehabilitation processes (cf. President's Council Report 2003; Hogle 2005). By granting individuals extra qualities in spite of their acquired disabilities, the rehabilitation of body functions and images may go beyond what is considered therapeutic. Nevertheless, despite this potential for 'enhancement', prostheses and other artificial rehabilitative technologies act as constant reminders of the initial loss of identity and pre-morbidity functions, as individuals either experience inconvenience using them or through them relive their pre-rehabilitation suffering. Rehabilitation processes of cancer victims through available technologies often follow therapy. This may fall on the borderland between therapy and enhancement; where the latter concept implies an intervention to improve human form and function after attention to genuine medical needs is no longer the focus of the attempts to restore sufferers' lives (cf. Degrazia 2005). Cancer rehabilitation is therefore a dynamic, ongoing health-orientated process designed to promote maximum levels of functioning related to health problems among the patients and is an ethical commitment by cancer care providers (Watson 1992).

Challenges to cancer rehabilitation in developed countries present differently from those in resource-poor countries, such as Kenya. The variation relates to how effectively cancer sufferers and their caregivers in both contexts are able to utilize various approaches to cancer rehabilitation to restore health and meet the needs of affected lives comprehensively. The cancer survivor's rehabilitation needs to reach far beyond the physical impairments from the cancer and its treatment (Mikkelsen et al. 2008, p. 216). However, cancer rehabilitation initiatives in both developed and developing countries may unwittingly fail to consider this reality. Four approaches are important for the rehabilitation of cancer patients at different stages of their treatment and care. The approaches entail preventive, restorative, supportive, and palliative rehabilitation (Dietz 1974; O'Toole and Golden 1991). Preventive rehabilitation should target reduction of predictable disability, through interventions such as implantation of an artificial bowel sphincter, physiotherapy, or appropriate training after therapeutic interventions. This may be provided by management in acute care hospitals or through outpatient services. Restorative rehabilitation implies efforts to help the patients and survivors to return to the state of perceived normalcy, which they experienced before cancer diagnosis and effects of radical treatment interventions. Rehabilitation technology at this stage may include an orthosis – orthopedic technology designed to support affected people, or correct deformity and improve function of movable parts of the body – or a prosthesis (O'Toole and Golden 1991).

The supportive rehabilitation approach follows the control of cancer, in spite of incomplete remission of the disease. In this case, patients are expected to endure treatment and general management of the disease through the use of some therapies. They may resume day-to-day activities as they visit the hospital for regular clinic reviews and control of progression and spread of the disease. Rehabilitation at this stage targets control of social and emotional impairment and improvement of physical functioning. Finally, palliative rehabilitation is necessary when the disease is advanced and the disability or impairment cannot be corrected. Exercise and counseling become crucial at this juncture to enhance well-being and existential strength. This is the hospice level of rehabilitation, which does not require active physical rehabilitation, but cognitive, emotional, and spiritual enhancement.

Despite the foregoing approaches to amelioration of cancer survivors' well-being, rehabilitation has rarely been discussed in the context of enhancement. But in a rehabilitation context, if one looks beyond the 'species-typical' scenario of the discourse on this theme to also consider individual experiences, and reflect about these experiences as an indication of what is 'normal', the question of where therapy ends and enhancement begins emerges prominently. This chapter examines the anthropological and moral questions that arise in this context, drawing on an ethnographic study that I conducted in a large public national referral hospital in Kenya. The chapter explores the implications of prostheses and other artificial supports for rehabilitation and the restoration of the bodies and lives of cancer patients and survivors. It examines cancer patients and survivors' perspectives on body and social function and identity that result from their experiences of rehabilitation, and explores the significance of rehabilitation as an aspect of enhancement. The chapter draws on Kenyan ethnography to discuss the adequacy of rehabilitation resources for restoration and enhancement of cancer victims' lives. It addresses the question: to what extend do available rehabilitation approaches address emotional and cultural issues regarding limbs, obliterated or impaired body organs, social and physical capacities, and prostheses? The central proposition is that cancer treatment and rehabilitation resources, such as prostheses, are not simply assistive, but can also be negative and continuous reminders of loss, which conceptualizations of rehabilitation from the enhancement perspectives can mediate.

## Ethnography of rehabilitation of cancer victims in Kenya

The ethnography was conducted in Kenyatta National Hospital (KNH) in Nairobi, between July 2005 and August 2006. I had return visits in 2007 and additional follow-up interviews with key informants in July 2010. KNH is the largest national referral and teaching hospital in Kenya. It caters for the treatment of referral cases from all over the country and new patients from Nairobi and neighboring districts. The adult cancer ward that accommodates both male and female patients was the main site of the study.[1] Other important study sites were the cancer treatment center review clinic and the radiotherapy clinic. The cancer ward admitted patients with head and neck, breast, cervical, colon, colorectal, prostate, esophageal, and gastric cancers. The hematology and obstetrics and gynecology departments manage the remaining types of cancer. Two separate wards offered pediatric oncology services, and one dealt specifically with retinoblastoma (cancer affecting the retina).

The incidence of cancer in Kenya is increasing rapidly. According to the cancer incidence report of cases recorded in Nairobi the five most common types of cancer among men in order of incidence are esophageal, prostate, non-Hodgkin lymphoma, liver, and stomach cancer (Sansom and Mutuma 2002). Cervical and breast cancers are the most common, among women, followed by ovarian, non-Hodgkin, and stomach cancers (Mutuma and Ruggut-Korir 2006; Sansom and Mutuma 2002). In Kenya, cancer is the third cause of death after infectious and cardiovascular diseases. The annual incidence and annual mortality due to the disease are estimated to be about 28,000 and over 22,000 cases respectively (Republic of Kenya 2010). Increasing incidence of cancer coupled with inadequate resources in the local healthcare system contribute to bleak prospects of improving the quality of life of cancer patients in Kenya.

Forty-two patients in the cancer ward, including one survivor of bilateral retinoblastoma (affecting the retinas of both eyes), were the main respondents in the initial 12-month hospital ethnography. Ten of the 42 main respondents had undergone either mastectomy or limb amputations and were using prostheses, with the exception of one male breast cancer patient. Most of the amputation and mastectomy survivors had their surgical operations elsewhere before referral for further radiotherapy and chemotherapy at KNH. The ethnography entailed direct non-participant observation and focused informal conversations with respondents on multiple occasions. The first part of the data collection entailed conversations with the purposively selected 42 inpatients. The study also included conversations with a key informant who participated in the Kenya Retinoblastoma Management Strategy project, 11 patients' relatives, 3 doctors, and 11 nurses. The second part of the ethnography focused on follow-up visits to ten purposively selected patients outside the hospital and at home to explore how they coped with the disease after or in-between hospital treatment sessions. Conversations with several other key informants and patients inside and outside the hospital supplemented the data. Interviews and informal conversations were conducted in either Kiswahili or English, respectively the national and official languages in Kenya. Transcriptions of conversations and interviews in Kiswahili were translated into English and analyzed together with those that did not require translations.[2] The ethnography demonstrates the ambiguity of patients and survivors' experiences of cancer treatment and rehabilitation. It reiterates the author's proposition that rehabilitation technologies may not simply improve cancer victims' quality of life but may also increase concerns about the sense of loss of former selves and well-being, which should be addressed.

## Quest for restoration of cancer sufferers' lives

Cancer therapy may have radical consequences, such as mastectomy (removal of one or both breasts), colostomy,[3] loss of eyesight, limbs, hair, or other body parts, and unusual skin pigmentation. These effects trigger physical and psychological suffering especially when patients' lived experiences contradict the expectation that hospital interventions such as intravenous feeding lines, under seal drainage tubes, or a wheelchair would be either temporary or restore daily life body functions. A patient's goals for participation in rehabilitation initiatives may differ from those of significant others or the professional treatment team (DeLisa 2001). The ethnography revealed differences in perspectives on cancer management outcomes among the patients or survivors and hospital caregivers, which influenced patients' satisfaction with rehabilitation initiatives. Hospital treatment and post-treatment rehabilitation experiences entail additional burden of mental suffering, which negates biotechnological efforts at restoration of health and quality of life among cancer sufferers. Even with extensive rehabilitation efforts, many cancer patients in the Kenyan ethnography, as in other resource-poor countries, were unable to effectively resume their normal life.

Most cancer patients in Kenya arrive at the hospital in relatively advanced stages of the disease. This contributes to the reality that recovery for most cancer patients often remains incomplete (Becker and Kaufman 1995), as restoration of normality is in essence unachievable. The disease and its treatment outcomes contribute to increasing social isolation and restricted autonomy in daily life. A cancer patient's life may therefore become a struggle. A 21-year-old amputee suffering from osteosarcoma, in a prosthetic leg, for example complained,

> that doctor came and told me that you have cancer... She came and told me, we will cut your leg from here [demonstrates and laughs mirthlessly]... She came with papers... and told me, we shall cut your leg here... So I was shocked... I asked her what do you mean

> and she told me, 'I am serious!' Even as I went to theatre, I thought she [the doctor] was joking, but when I woke up I found she was serious. I found my leg had gone... She is the one who told me that I had cancer, just bluntly like that... I have been told that I will have to lose my hair, I have to vomit, diarrhoea, and things like that. So far I have lost my hair. But I decided to shave the remaining, but I have been told that more I will regain the hair.
>
> (Mr Kamenju, 21 years, interviewed; January 2006)

Temporary or permanent body alterations worsen the physical and emotional experience of cancer, as with other stigmatizing diseases (cf. Fife and Wright 2000; Gilmore and Somerville 1994). Stigma in such cases relate to feelings of loss of social and individual status and identity. In the minds of 'normal' people, those who are stigmatized are seen as persons whose identities are spoilt and tainted, rather than whole and socially acceptable people (Carricaburu and Pierret 1995; Cumming and Cumming 1965; Goffman 1963). Loss of identity is therefore a fundamental aspect of perceived losses among cancer patients, especially if they experience significant disfigurement or damage of the body image and functioning due to the disease and treatment. Physical, mechanical, and/or aesthetic losses have implications for emotional and social well-being, as they also affect a person's social image and function. This experience restricts cancer victims' social space to local and familiar territories where questions and curiosities of acquaintances and strangers are minimal.

The experience of head and neck cancer affect patients in multiple ways. The disease and treatment may distort appearance, expression, and speech, which make communication difficult; the performance of such normal activities as swallowing, blowing the nose, or even breathing may be impeded; and the patient's very livelihood may be jeopardized, either by his disability or others' reaction to it (Larsen 1982, p. 119). This experience further affects a patient's sense of self-worth and identity and calls for rehabilitative and supportive measures even before the commencement of treatment. Consider the example of a nasopharyngeal carcinoma patient's observation:

> people were wondering what was going on, or 'what kind of disease does she have?'... because when I was being 'burnt' (undergoing radiotherapy) my appearance changed and I became black, and even people were wondering 'aye! What is wrong with this one?' I used not to eat, but just taking milk. So you know when I started being burnt

> there was a hole which developed... here (in the nose), you can even hear the way I am talking now. When I talked to people they could not hear me... I could not even eat at that time.
>
> (Ms Keberia, 37 years, 16 March 2006)

The above excerpt points to the need for some form of enhancement before treatment and after rehabilitation – beyond therapy – in cancer management. As Cheville (2005) points out, it is imperative to have cancer rehabilitation as an integral part of the care plan and this be given equal consideration while decisions about surgery, radiation, and chemotherapy are made. Important aspects of the rehabilitation plan should include emotional and psychosocial dimensions of cancer and therapy side effects.

Chronic illness and altered physical characteristics introduce sufferers to new life experiences. Healthy 'normal' people, however, tend to influence the consciousness of cancer sufferers regarding their 'deviation' from 'normal life'. 'Normal people' may remind cancer sufferers of their new identities through open curiosity about their bodies or the use of stigmatizing labels. Psychological support and the technologies used to bring them out of such stigmatizing states can constitute 'enhancements', according to the definition given by the United States President's Council on Bioethics (President's Council Report 2003), as they are initiatives that go beyond therapy and mere restoration of biological and physical body functions. While the technologies and artifacts used to make the lives of cancer patients and survivors better are rehabilitative, they ultimately contribute to the reinforcement of new personal identities and experiences. However, cancer patients and survivors may perceive rehabilitative artifacts and people's curious reaction to their new identities as tokens of loss, especially when they reinforce lived experiences of physical and social impairment.

Supportive rehabilitation is important for enhancement of personal coping potential in the face of radical cancer treatment and its outcomes. Cancer *per se* ushers victims into a new life world; into the 'kingdom of the ill' (cf. Sontag 2001) or the 'village of the sick' (Stoller 2004). Rehabilitation technologies such as prostheses would be tokens of gain if they become part of the metaphors of empowerment and improved quality of life. These images would facilitate restoration and enhancement of victims' abilities to face physical and psychosocial consequences of the disease and its management at personal and social levels (Sontag 2001; Van Dongen 2008). The loss that cancer victims experience is both physical and psychosocial; hence rehabilitation should target comprehensive restoration of disrupted lives. Patients and survivors need adequate opportunity to grieve their loss to the point of closure. The ethnography in Kenya pointed to the fact that cancer victims experience further loss in clinical settings when caregivers overlook the need for 'holistic' rehabilitation and restoration of both physical and non-physical lives. Somatization or medicalization of loss during rehabilitation processes blocks the holistic healing that would enhance sick bodies and psychosocial and emotional impairment. Clinical professionals on Kenyatta hospital's cancer ward on occasion resorted to the use of sedatives and antidepressants to relieve sufferers' psychological pain and anxiety due to loss of functional abilities. Such medications give only transitory support by interfering with the process of grieving (cf. Maguire and Parkes 1998) and delaying closure, yet grieving and closure are necessary for comprehensive enhancement of cancer victims' quality of life.

## Coping with loss of body parts

Drastic cancer treatment involving procedures such as limb amputation, enucleation (removal of the eyeball),[4] unilateral or bilateral mastectomy, and colostomy entail perceived untimely loss. Most of the respondents and key informants indicated that cancer patients who underwent surgical removal of organs were concerned about the 'whereabouts of the removed body part'. When patients were not well-prepared psychologically for surgical removals, they attributed their loss to erroneous medical decision-making. They often had difficulty in believing the fact of their loss. Such mistrust characterizes the behavior of cancer victims who tend to avoid visible reminders of their suffering (Maguire and Parkes 1998) and who express anger and bitterness about some treatment and rehabilitation processes. Cancer victims in the Kenyan ethnography perceived greater loss when medical personnel failed to explain possible outcomes of available alternative treatments and rehabilitation technologies. Physicians' and surgeons' ability to clearly demonstrate that the removed part was 'spoilt' and a threat to the rest of the body allayed bitterness about what patients perceived as undeserved or premature loss. Such feelings were often associated with perspectives on the loss of physical attractiveness (body image), loss of occupational and other physical functions, or both together. Surgical treatment, chemotherapy, and radiotherapy result in perceived identity and personhood loss. This relates to a new and relatively frail biographic experience, which in turn shape perceived loss of self-esteem, grief, and depression (Maguire and Parkes 1998).

Many people expect that medical treatment and artificial body supports, together with physical and psychological rehabilitation, can enable cancer patients and their families to come to terms with their altered selves and new lives. However, as aforementioned, prostheses and other rehabilitation resources may further trigger memories of loss. Some users of rehabilitation artifacts may perceive them as either not good enough and cumbersome, or causing additional pain or stigma. Perceived loss of pre-diagnosis or pre-therapy identities, selves, and livelihood security, and inadequate rehabilitation processes and facilities exacerbates the situation. While it is true that prostheses and other rehabilitation technologies can contribute positively to the quality of life of cancer sufferers, it is also true that these objects constitute daily tokens of otherness; a consequence of perceived loss of the authentic self. They attract public attention to the users' difference, and constitute the focal symbols for overt or covert elicitation of sympathy. Social and cultural notions of what it means to be fully human – a notion which I address in more detail below – further aggravate the stigma that victims of cancer and others associate with prostheses and rehabilitation objects and techniques. Attempts to rehabilitate and restore the lives of cancer patients may reinforce notions of lost social and physical identities, and sustain silent grief over 'social death'. Cancer patients and survivors for instance often struggle in vain to restore their gender identity, sexuality, and perceived loss of physical integrity (Manderson 1999).

## Grieving the loss and 'being incomplete'

Grief about loss due to disease or therapeutic intervention can reach closure through appropriate grieving rituals, which vary with social and cultural contexts. Religious rituals and everyday life practices may interact with rehabilitative technologies and approaches to bring about a more effective restoration of social and emotional life. For example, the hospital ethnography in Kenya revealed that amputees and their kin often wish to have the surgically removed parts at their own disposal. This may be followed by private or public ritual – such as burial of the amputated limb or breast and prayers – to rationalize the loss and experience a transition to closure. When such processes take place, the physical loss may not translate to spiritual loss, and the wearing of a prosthesis or artificial body part may not protract the feeling of physical and social loss.

The psychological reactions to biophysiological and social losses due to disease or treatment constitute a category of grief that requires spiritual enhancement to effect closure. Yet, the techniques applied in the rehabilitation of cancer patients and survivors contribute to ambivalent feelings, as they have the potential to both alleviate and increase existential suffering. Cultural beliefs and attitudes, as well as the quality of information, advice, and emotional support available to patients and their families shape the construction of the severity of the loss and the efficacy of rehabilitation strategies. The cancer patient or survivor therefore requires further support from relevant healthcare personnel after discharge and continuous monitoring of his or her rehabilitation needs. The family, on the other hand, need information about how the cancer could affect the patient's and his or her peers' psychosocial situation and information about support during and after treatment (Mikkelsen et al. 2008, p. 218).

A majority of cancer patients in the Kenyan ethnography expressed unresolved anxiety about loss of body parts to cancer and treatment. A common metaphor that most of them used was of 'being half dead'. Being physically incomplete at death for many Kenyan, as in other African cultures, portends spiritual ambiguity; death while lacking some or all of a body part causes fear of spiritual condemnation of the dead person and his or her family. In this regard, treatment and rehabilitation approaches call for the spiritual or psychosocial enhancement dimension. Cancer patients suffer many functional deficits that result in long-lasting emotional and social distress, which reduces their quality of life. The concept of quality of life entails physical, emotional, social, and cognitive function, which may be positively influenced by physical exercise (Fialka-Moser et al. 2003, p. 18).

Stigmatizing references to prostheses reflect the perceived ambiguity and liminality of a prosthesis wearer's personhood. Loss of a body part on the other hand may imply supposed permanent loss of normal functioning in the long run. Experience of these phenomena aggravates the struggle to overcome worries about the ominous idiom of 'being half dead'. Beliefs of some indigenous ethnic groups in Kenya depict the removal of a body part due to an accident or disease as 'an early phase of death'. The 'burial' of a part of the body marks the beginning of life as an incomplete physical and social being. This accounts for reported kin resistance to therapeutic removal of retinoblastoma victims' eyes. Reduced social interaction in the case of limb amputees, mastectomy survivors, and people who have lost their eyes, epitomizes ominous processes that result in relegation of sufferers to social invisibility, as the dead. The loss of body parts means a major disruption: the loss of both physical health and social life. This perceived social death is due in part to the idea of the 'partial physical death' that is attached to such people.

A nurse observed, for instance, that affected people feel that 'someone is no longer a complete woman after a mastectomy'. Cancer and its treatment may indeed disfigure people and impose a heavy psychological burden; people stigmatize survivors of breast cancer, for instance, and their spouses are inclined to reject them (Fialka-Moser et al. 2003).

Grieving processes are significant in the rehabilitation of cancer patients who lose body parts. Loss and grieving are both individual and social. Disfigurement, functional impairment, and attempts to restore well-being of cancer patients involve people who take part in therapy management, including patients, their families and associates, or the therapy management group (Janzen 1987; Larsen 1982, p. 119). Use of rehabilitation technologies without attention to the emotional components of healing is therefore insufficient as a way of restoring the health of cancer survivors. There are similarities between grief at the loss of body parts and grief caused by the death of a loved one (cf. Parkes 1975; Maguire and Parkes 1998). The emotional pain that bereaved people experience at the time of their loss can last for a long time afterwards, as subsequent memories and reminders of the lost loved one may continue to cause emotional distress.

Cancer rehabilitation artifacts and technologies may exacerbate and prolong emotional suffering, since they act as tokens of the loss, in particular if the person has not had adequate opportunity to explore and address related existential questions in order to complete the grieving processes. The flow of appropriate information, educational messages, and counseling tailored to the unique needs of individuals may alleviate the physical and emotional suffering that results from the disease, therapy, and rehabilitation. However, attempts at restoring cancer patients' physical lives in Kenya are often lacking in strategies to heal disfigured emotional and social lives. Successful adjustment to mastectomy, enucleation, and limb amputation and the related prostheses, for instance, is less likely as levels of pain and social isolation increase (Williamson et al. 1994). Therefore, cancer patients and survivors crave not only rehabilitation of their disabled condition but also restoration and enhancement of their former self, of their disrupted biographies and social relationships (Bury 1982), and of their damaged body or self-image.

## Perspectives on mastectomy and breast prostheses

While breast prostheses or improvised props are designed to improve self-image and emotional stability, they may on occasion make the affected women feel even more conspicuous. One respondent, noted that breast prostheses often reminded her of the loss she experienced, as 'they are hardly replacements of natural breasts'. Side effects of breast cancer treatment, especially mastectomy, have dire implications for cancer survivors in Kenya as in other parts of the world. Therefore, the recent introduction of new breast-prosthetic products was aimed at helping breast cancer survivors reclaim their image, self-esteem, and confidence, which they lost during the disease recovery (Githinji 2012). However, the visibility of external breast prostheses and brassieres may reduce self-esteem among users, in the same manner as conspicuous cancer treatment side effects such as hair loss. Key informants who participated in the hospital ethnography noted that breast-prostheses users were very cautious about the public attention they attracted due to their altered physical (self-) images, which also implied perceived decline in social worth and a new phase of psychological suffering. Some patients and survivors tried to gain emotional and physical confidence by using external breast prostheses to restore a culturally defined traditional womanly shape (Figure 6.1).

External prostheses are more accessible in Kenya than breast implants as the latter involve more expensive technology. Furthermore, poor women who cannot access proper external breast prostheses have to make do with improvised breast props. The external prostheses or improvised augmentations, however, fail to facilitate women's quest to rehabilitate and restore their desired womanly shape and feminine appearance for at least three reasons. First, wearers of breast prostheses may feel conspicuous and uneasy due to their consciousness of the

*Figure 6.1* A pair of Discrene brand mastectomy breast forms with optional false nipples glued on (taken by Alice Markham, 2002, Source: Wikicommons)

artificiality of their new image. Second, available prostheses never replace the natural body in domestic relations such as spousal intimacy. Third, the fact that many people may know one to be a mastectomy survivor elicits mistrust regarding claims of restored identity and social reintegration. These factors influence affected women's attempts to gain emotional and physical confidence by using external breast prostheses to restore a culturally defined feminine body shape. In addition, women who sought breast prostheses soon after treatment complained of skin irritation and dearth of preferable sizes and colors. Financial inaccessibility and aesthetic unattractiveness of available breast prosthesis constrained their utilization among needy cancer victims in Kenya. Most of the prostheses supplied were pink and of the style shown in Figure 6.1. The diversification of the supply of breast prostheses in the world market today to include fashionable colors – black, white, off-white, nude, cognac – provides women with wider choice. This makes it possible for users to choose among colors that are ideal for most complexions and are handy to wear underneath most outfits (Githinji 2012).

Cancer patients and survivors who are lucky enough to get external breast prostheses feel socially more secure in spite of the aforementioned limitations. The prostheses contribute to the partial restoration of their self-esteem through improvement of their impaired bodies and restoration of their 'woman image'. Some breast-prostheses users indicated that they had regained the confidence to overcome public suspicion and stigma, and avoid being the subject of overt or subtle gossip. For other breast cancer survivors, however, prostheses are not fully enhancing, as negative feelings about them linger. These feelings reflect the experience of having a disrupted social body and biography (Bury 1982; Scheper-Hughes and Lock 1987; Synnott 1993), as cancer and the prostheses limit participation in socially valued gender-based activities. Some participants in the ethnography in Kenya harbored negative sentiments about prostheses, describing them as 'putting on something that is not yours', which made them feel inadequate.

While an artificial breast becomes an *attached* body part, the wearer may still feel deficient as it is not fully *their* body part. This feeling may be more pronounced among women who use prosthetic brassieres for aesthetic enhancement of their natural breasts than among victims of mastectomy. One respondent observed that the feeling of deficiency in spite of using prosthetic breasts has a negative influence on marital relationship, because 'the crucial part of their gendered or sexual body has been removed'. Breast prosthesis may thus be a token of the initial loss, which leads to further losses such as broken marriages and families and the disruption of entire lives. Limb prostheses may restore survivors' ability to participate in social rituals, such as romantic and sexual relationships (cf. Murray 2005), but breast prostheses can inhibit this. Karakasidou (2009, p. 108) equates the use of breast prostheses to the reconstruction and concealment of cancer, through which physicians and patients revive the 'conspiracy of silence' by rendering the 'face' of cancer socially normal and aesthetically beautiful. More significantly, use of prostheses can turn cancer into a cosmetic issue, and this silences debate about the cause of the scar and the cancer (Ibid., p. 102). This constitutes a loss of the opportunity for the expression of subjective experience and recognition of the need for further emotional support after cancer and surgical tribulations.

While a prosthesis may improve body image, it can also remind users of their bad experiences and stifle their quests to build up much needed support networks. According to a radiotherapy clinic nurse, some of the women would say, 'I do not want to have anything artificial on me...it reminds me how I have suffered...and memories of what has happened to me in my life'. Acceptance of breast prostheses in Kenyan society is therefore fraught with ambiguity. Similarly, replacement of a body part may not completely preclude the private and public perceptions of removal, loss, and replacement as something sinister. This further shapes the visibility of the disease; the stigma attached to it and the perceived sense of loss. The effects of the cancer site and the treatment of the disease have negative implications on everyday social lives and livelihood, which aggravate the impact of perceived loss (Warren and Manderson 2008). Three dimensions of cancer patients' anxiety about their bodies emerged from narratives about breast cancer and prosthesis use. First, radical cancer treatment causes indelible scars and severe alteration to or loss of some body parts. This quintessentially shaped the disruption of self-perception as the patients convalesced, rehabilitated, and adjusted to the (usually) permanent physical reminders of the surgery (cf. Manderson 1999, p. 381). The second dimension relates to the fact that cancer and its treatment weaken both the body's natural strength and resilience. Efforts to restore day-to-day health and body vitality in the context of cancer therapy may therefore require enhancement through nutritional and pharmaceutical supplements. However, while the use of dietary supplements and medication can promote more normal body functioning during therapy and recovery, they also constitute tokens of loss; that is, loss of the capacity to rely on ordinary food and body immunity for ordinary body defense and healing. Efforts to boost natural body functioning through medical technology, however, often come up against obstacles such as inadequate resources and psychological adjustment to the cancer treatment and rehabilitation interventions.

The third dimension of cancer patients' fear involved treatment side effects that are perceived to alter body image and affect individual identity and self-esteem. This fear partly accounted for direct or concealed resistance to initial surgical operations and subsequent therapies. Resistance to interventions involving loss of a body part resulted from a general mistrust of doctors. Many patients and their families initially refused radical interventions such as amputation, mastectomy, and enucleation, doubting the accuracy of the doctors' decisions and interventions.

## Prosthetic eyes: Loss and gain

Key informants narrated the ordeals that accompany the acquisition and use of synthetic eyes. They reported that removal of eyes and their replacement with artificial ones is both painful and emotionally traumatizing. Removal of children's eyes must be determined not only between doctors and patients, but also between doctors and the children's guardians or kin. The process of eye replacement with a prosthesis is relatively less complicated for children, when their parents and relatives consent, because children adjust better, physically and socially, to living with just one functional eye. Delaying the use of prosthetic eyes until adulthood makes the experience extremely painful due to the process of growth and healing.

One informant indicated that poorly fitted eye prostheses that fall out of the eye socket, or that distort the aesthetic effect of the face cause additional suffering to the wearers and their caregivers, while removing and fixing eye prostheses worsens the painful experience. This constitutes the negative consequences of what one informant described as the 'pain of giving the wrong eye'. However, most informants, including a retinoblastoma survivor, appreciated the fact that replacement of an eye with prosthesis improved the facial image of children and adults alike. Such beneficiaries would appreciate this directly by looking at themselves in the mirror and by observing other people's reactions. When the constraints on the removal of a child's eye are overcome, eye prostheses are appreciated, in principle, as the beneficiaries grow up. However, this depends on whether they are available and affordable. An artificial eye generally reduces stigma, particularly when only one eye is affected. A retinoblastoma survivor, observed:

> It makes you look more of the original person. It is more useful when very few people notice that it is not a natural eye; people are suspicious when you lack one eye. Nobody may believe a person without one eye or both was sick. Some people think maybe the person may have been a thief then the eye was poked out... The artificial eye makes people accept us.

The stigma of lacking one or both eyes is similar to that experienced by those who wear limb or breast prostheses, though in addition there is the particular association of the loss of eyes with evil or misconduct that is prevalent in local discourse in Kenya. Retinoblastoma victims, like other people who lose body parts due to disease, would prefer an artificial replacement, even if the prosthetic eye is obviously not functional.

Imagined sources of artificial eyes, however, soon becomes a problem to users. One informant confirmed the idea, popular in some Kenyan societies, that people liken prosthetic eyes to either 'sheep or goat eyes'. This affects the wearer's self-esteem and reduces the chances of people accepting therapeutic eye removal and replacement with prostheses. Most informants indicated that, in social and domestic interactions, artificial eyes attract condescending attitudes towards cancer survivors and their kin. A nurse observed that at times some people demeaned eye cancer survivors who used prostheses, referring to them as people 'wearing animal eyes'. In addition, a retinoblastoma survivor noted that replacement of both eyes with eye prostheses or the use of dark glasses to conceal absence of eyes would limit the person's chances of being assisted by members of the public, for example to cross a busy street. The public, particularly in large urban areas, are often cautious of 'being conned' by individuals who fake a disability, such as those who pose as blind. Use of prostheses or dark glasses in this sense would contribute to the loss of opportunities for support and assistance for people with genuine disabilities.

Local witchcraft and magic beliefs in Kenya further contribute to misgivings about prostheses. An artificial eye or limb may cause curiosity and even suspicion, particularly when their colors are not similar to the wearer's real complexion. More significantly, when people consider a person 'normal' or of outstanding personality, the discovery that he or she uses a prosthesis might create an aura of mystery around the person.

Prostheses, such as an artificial eye or tooth, may be associated with objects of mystical power that contribute to a perceived personality or character peculiarity. Similarly, the fact that an eye prosthesis, like other synthetic body parts, can be removed and replaced at will to enchant, elicit sympathy, or inspire awe in an audience, shape further the perceived ambiguity of prostheses. The mystery of such body rehabilitation technologies may thus create grounds for speculation about the perceived use or misuse of mystical or extraordinary abilities inherent in artificial body parts. For retinoblastoma survivors, therefore, having artificial eyes is both a token of loss and gain. In the first place, it may be used for aesthetic purposes. It improves the survivor's self-image and self-esteem. Nevertheless, this positive outcome depends on the social and cultural environment in which people define normal personhood and markers of abilities or disabilities.

## Prosthetic limbs

A middle-aged woman suffering from fibrosarcoma[5] had resisted amputation of her leg for over two months. As she adjusted to her new leg-amputee identity, she regretted the loss of what she still thought might have been a possibly viable leg. Prosthetic limbs used in Kenya essentially constitute a plastic drainpipe with a rubber foot or arm attached. The artificial limbs provide renewed hope for many victims of disease and accidents. However, amputees often struggle to get used to their artificial limbs. Most informants confirmed that users of artificial limbs had to adjust to the experience of their new limbs not being as good as the ones they had lost. Some informants, when they commenced using crutches and artificial limbs, complained of them as being either too heavy or of an inappropriate size. Nevertheless, some amputees and other physically disabled people in Kenya prefer crutches, sticks, and artificial legs to wheelchairs. For most of these informants, a wheelchair would facilitate movement but constrain personal freedom.

A 25-year-old respondent said that he had gradually gotten used to his artificial leg. His late grandfather's prosthetic leg had been preserved in their house, reminding his family of a problem he had in common with his deceased grandfather. The leg prosthesis for him, however, embodied the loss of school time, livelihood, and esteem in his lineage. Since the amputation and the acquisition of an artificial leg, he recalled, his father had also lost working time when taking him for treatment and helping him practice with his new leg. Constrained movement when using prosthetic limbs, crutches, and wheelchairs leads to lost sociability;

younger people in particular drift into isolation and loneliness. A 24-year-old female patient, for instance, grappled with accepting her new amputee status. She was concerned that after cancer treatment, amputation, and hospitalization, it would be difficult to restore her social life and sense of self-worth. Restricted movement and reduced interaction with and even loss of friends worsened her pain. She missed her final high school examination because of the illness, hospitalization, and practicing using her crutches. Like other amputees who use crutches, grieving the loss of both her limb and her ease of movement lingered long after the amputation.

Some cancer victims who had acquired artificial limbs felt that they should always have them on, even when they went to bed at night. As in the case of some users of breast prostheses, artificial limb amputees would anticipate that the prosthesis would sufficiently replace their lost limb. Conversely, the fact that artificial limbs can be detached means that some users would have the chance to display them as evidence of their loss and disability in order to attract sympathy and help, as in major urban areas. The fact that artificial limbs can be removed and attached at will contributes to the simultaneous feelings of loss and gain associated with them. This reveals the philosophical foundations of rehabilitation. Rehabilitation does not imply cure alone, but also addresses the need to focus on the long-term recovery of, or adjustment to, functional losses (cf. Ory and Williams 1989, p. 67). New users struggle to adapt to rehabilitation resources, with little or no physiotherapy and psychosocial support. The goal of rehabilitative therapies should be to enable persons to maximize (or enhance) the functions they have, and 'to retrain them to modify their behavior to compensate for those they have lost' (Becker and Kaufman 1995). Rehabilitation efforts should transform prostheses from being burdensome reminders of loss to technologies for the restoration of quality of life. The culture-specific construction of the qualities of prostheses and other rehabilitation technologies presents a particular challenge. For example, as with artificial eyes, the color of artificial limbs matters to users. Informants noted that people become more suspicious when they see color disparities between the prostheses and their bodies, as such disparities contribute to the prevailing doubts and superstitions in the discourse about the source of prostheses. A nurse's observations on misconceptions about the sources of prostheses are indicated in the extract below:

> they think that maybe doctors removed the artificial eyes from dead people and modified them into prostheses, especially if the color is

too different from the complexion of the users. Prosthetic limbs, fingers, and synthetic hair need to be of acceptable colors and should not be too different from the complexion of the users.

## Cosmetics and synthetic hair

Loss of hair and changes in skin pigmentation also featured in cancer survivors' short- and long-term concerns about rehabilitation resources. The results of the ethnography highlight ambivalence about the use of cosmetics and synthetic hair to disguise the effects of cancer and treatment. Body surfaces affected by cancer may turn dark or black due to chemotherapy and radiotherapy, and these dark patches may spread from the radiotherapy sites. Further negative effects and emotional suffering may result from the use of cosmetics to rehabilitate body image and restore health, as they can lead to skin irritation or uneven results from bleaching. Informants told of body creams and cosmetics that caused darker and lighter patches concurrently, which attracted curiosity, stigma, and gossip. Bleaching of the skin as a cosmetic treatment is generally seen as desirable by many women in Kenya, to lighten their skin. However, the poor quality of cosmetics and improper use often cause undesirable side effects. These negative consequences of cosmetics can be understood as an aspect of failed rehabilitation and body enhancement efforts, which some people blame on victims themselves. The cancer survivors, in this sense, endure a double loss: both of their 'natural' body image, and of their hopes about the restorative effects of cosmetics.

Hair loss can be particularly problematic for cancer sufferers, for it parallels notions of loss of individuality, attractiveness, and status, especially among women (Hansen 2007). As in the case of mastectomy, hair loss is associated with the loss of a key attribute of womanhood, and thus with loss of femininity and sexuality (Bachelor 2001). This loss affects women's social relations and daily lives significantly. Furthermore, the fact that hair loss (and some skin-related difficulties) are among the symptoms of other diseases, especially HIV/AIDS, which are stigmatized, can also be problematic. Cancer victims may thus feel doubly stigmatized. Thinning of hair and change of hair color foretell hair loss, for which men and women seek different remedies. Kenyan men hardly ever use artificial hair (wigs) when hair loss sets in. They instead prefer to 'shave clean'. Women, on the other hand, do prefer to supplement their lost hair. The use of hair additions in Kenyan societies is conventionally a female practice aimed at enhancing or restoring beauty and self-image, even during good health. However, synthetic hair – as opposed to real human hair – is, in normal circumstances, neither appreciated nor preferred, and if a woman uses synthetic hair during or after cancer treatment she may elicit suspicion. Just as with artificial limbs and prosthetic eyes, some people have qualms about the source of artificial hair. Study participants reported that some people fear that artificial hair might have been 'taken from the deceased, or horses' tails'.

Reactions to cosmetic treatment and artificial hair in popular discourse, rehabilitation structures, and social relations influence how cancer patients and survivors view them – as either rehabilitation resources or a means of restoring well-being (cf. Rosman 2004). The use of artificial hair is an extension of culturally embedded signs, which should be taken into consideration in rehabilitation processes. Negative discourses about synthetic or human hairpieces influence their acceptability and their role in the process of recovery and healing. Rehabilitation of cancer victims therefore requires a careful consideration of the challenges to processes of normalization and integration of the individual in the personal, social, and cultural contexts of their experiences.

## Conclusions: Beyond restoration

Attempts to rehabilitate and restore bodies and lives affected by cancer have both physical and emotional dimensions. Users of prostheses and other artificial supports thus require not only physical but also social and emotional adaptation to their new conditions. The success of recovery and healing depends on how adequately and efficiently rehabilitation processes can restore cancer victims' bodies and lives to the state that cancer survivors enjoyed before diagnosis and treatment interventions. Cancer treatment disrupts potentially already fragile lives, particularly in resource-poor countries such as Kenya. Rehabilitation in such a context is set against a backdrop of pre-existing physical and well-being limitations, apart from those that the disease itself brings. Owing to this, rehabilitation initiatives need to focus on dealing with all the livelihood constraints that the disease, treatment interventions, and pre-diagnosis vulnerabilities impose on victims. The efforts at restoration of health and daily social functioning thus require additional initiatives to help survivors and patients complete their trajectories of grieving for their perceived losses and to arrive smoothly at emotional closure. Perceived losses that beg for both restoration and enhancement strategies extend to non-physical aspects of disrupted biographies and well-being.

While rehabilitation measures can restore some social and physical functions – albeit inadequately in some cases – they can also add new ones. These new functions include perceived additional abilities embodied in rehabilitation artifacts such as eye and limb prostheses, which may also elicit sympathy, curiosity (associated with freaks), or suspicion towards wearers. These perceived extra attributes of rehabilitative artifacts and technologies constitute unintended side effects of the intended functions of the prostheses (which are to look less unpleasant and restore health and socially accepted identities). While prosthesis wearers can exploit additional advantageous qualities resulting from rehabilitation processes for their own social and physical survival, some of the new functions also bring about additional worry and require new social and psychological support for coping. A disturbing appearance, for instance, might increase prosthesis wearers' body image anxiety. Conversely, some people may derive personal benefit from their ability to wear and remove prosthetic limbs or eyes at will, and from using aesthetically acceptable hair additions, cosmetics, and breast prostheses.

Initiatives for rehabilitation of cancer survivors in Kenya reflect a struggle to comprehensively salvage bodies and lives from stigmatizing states. Restoration of health in this sense does not imply cure and rehabilitation of the body alone. It must extend to efforts to address social, cultural, emotional, and spiritual needs related to the quest for enduring, revitalized lives and acquired and ordinary functional limitations. This process constitutes 'enhancement', because it involves initiatives that go beyond therapy and the mere reinstatement of biological and physical body functions. Inadvertently, some rehabilitative technologies and artifacts that are used to make the bodies and lives of cancer patients and survivors better contribute to reinforcing new personal identities and experiences. In the absence of strategies to resolve the emotional implications of diminished body functioning due to cancer and its treatment interventions, rehabilitation artifacts and technologies may actually serve as perpetual reminders of loss rather than of gain.

## Notes

1. The data on which this chapter is based are part of a larger ethnography for the author's PhD dissertation; see Mulemi, B.A. 2010. *Coping with Cancer and Adversity: Hospital Ethnography in Kenya*, Leiden: African Studies Centre. The hospital ethnography was approved by the hospital's Research Clearance and Ethics Committee.

2. The study entailed continuous pre-analysis of recurrent themes during the fieldwork. Data were transcribed and prepared for analysis using the Maximum Qualitative Data Analysis program. Frequently recurring themes relating to effects of experiences of cancer and its management on emotions and daily lives were noted. Analysis of notes from direct observation and informal conversations supplemented data from focused in-depth conversations. I use pseudonyms to refer to key informants in this chapter.
3. Colostomy is a surgical procedure which directs one end of the large intestine out through the abdominal wall. Fecal matter moving through the intestine drain into a bag attached to the abdomen.
4. Enucleation, done for intraocular tumor excision, is the removal of the eyeball, while leaving the adjacent structures of the eye socket and eyelids intact.
5. Fibrosarcoma is a malignant tumor that originates from fibrous tissue of the bone. The cancer spreads on long or flat bones such as femur, tibia, and mandible and is common among men and women of 30–40 years of age.

## References

Bachelor, D. (2001): Hair and cancer chemotherapy: Consequences and nursing care – A literature study, In: *European Journal of Cancer Care*, 10, pp. 147–163.

Becker, G. and Kaufman, S. R. (1995): Managing an uncertain illness trajectory in old age: Patients' and physicians' views of stroke. In: *Medical Anthropology Quarterly*, 9(2), pp. 165–187.

Bury, M. (1982): Chronic illness as a biographical disruption. In: *Sociology of Health and Illness*, 4, pp. 167–182.

Carricaburu, D. and Pierret, J. (1995): From biographical disruption to biographical reinforcement: The case of HIV-positive men. In: *Sociology of Health & Illness*, 17(1), pp. 65–88.

Cheville, A. L. (2005): Cancer rehabilitation. In: *Seminars in Oncology*, 32, pp. 219–224.

Cumming, J. and Cumming, E. (1965): On the stigma of mental illness. In: *Community Mental Health Journal*, 11, pp. 135–143.

DeGrazia, D. (2005): Enhancement technologies and human identity. In: *Journal of Medicine and Philosophy*, 30, pp. 261–283.

DeLisa, J. A. (2001): A history of cancer rehabilitation. In: *Cancer*, 92, pp. 970–974.

Dietz, J. H. (1974): Rehabilitation of the cancer patient: Its role in the scheme of comprehensive care. In: *Clinical Bulletin*, 4(3), pp. 104–107.

Fialka-Moser, V., Crevenna, R., Korpan, M. and Quittan, M. (2003): Cancer rehabilitation: particularly with aspects on physical impairments. In: *Journal of Rehabilitation Medicine*, 35, pp. 153–162.

Fife, B. and Wright, E. (2000): The dimensionality of stigma: A comparison of its impact on the self of persons with HIV/AIDS and cancer. In: *Journal of Health and Social Behavior*, 41(1), pp. 50–67.

Gilmore, N. and Somerville, M. A. (1994): Stigmatization, scapegoating, and discrimination in sexually transmitted diseases: Overcoming 'Them' and 'Us'. In: *Social Science and Medicine*, 39, pp. 1339–1358.

Githinji, P (2012): *Deacons Kenya Limited Unveils Products for Breast Cancer Survivors,* http://kehpca.org/deacons-kenya-limited-unveils-products-for-breast-cancer-survivors/, date accessed 20 October 2013.
Goffman, E. (1963): *Stigma: Notes on the Management of a Spoiled Identity.* Englewood Cliffs, NJ: Prentice Hall.
Hansen, H. P. (2007): Hair loss induced by chemotherapy: An anthropological study of women, cancer and rehabilitation. In: *Anthropology and Medicine,* 14(1), pp. 15–26.
Hogle, L. F. (2005): Enhancement technologies and the body. In: *Annual Review of Anthropology,* 34, pp. 695–716.
Janzen, J. M. (1987): Therapy management: Concept, reality, process. In: *Medical Anthropology Quarterly,* 1(1), pp. 68–84.
Karakasidou, A. (2009): Modern aesthetics and the cancerous body reconstructed. In: *Tijdschrift Medische Anthropologie,* 21(1), pp. 107–116.
Larsen, G. L. (1982): Rehabilitation for the patient with head and neck cancer. In: *The American Journal of Nursing,* 82(1), pp. 119–121.
Maguire, P. and Parkes, C. M. (1998): Coping with loss: Surgery and loss of body parts. In: *British Medical Journal,* 4(316), pp. 1086–1088.
Manderson, L. (1999): Gender, normality and post-surgical body. In: *Anthropology and Medicine,* 6(3), pp. 381–394.
Mikkelsen, T. H., Søndergaard, J., Jensen, A. B. and Olesen, F. (2008): Cancer rehabilitation: Psychosocial rehabilitation needs after discharge from hospital? A qualitative interview study. In: *Scandinavian Journal of Primary Health Care,* 26, pp. 216–221.
Murray, C. D. (2005): The social meanings of prosthesis use. In: *Journal of Health Psychology,* 10(3), pp. 425–441.
Mutuma, G. Z. and Ruggut-Korir, A. (2006): Cancer incidence report. Nairobi 2000–2002. In: *Nairobi Cancer Registry,* Kenya: Kenya Medical Research Institute.
Ory, M. G. and Williams, T. F. (1989): Rehabilitation: Small goals, sustained interventions. In: *The Annals of the American Academy of Political and Social Science,* 503(1), pp. 60–71.
O'Toole, D. M. and Golden, A. M. (1991): Evaluating cancer patients for rehabilitation potential. In: *The Western Journal of Medicine,* 155(4), pp. 384–387.
Parkes, C. M. (1975): Psycho-social transitions: Comparison between reactions to loss of a limb and loss of a spouse. In: *British Journal of Psychiatry,* 127, pp. 204–210.
President's Council on Bioethics (2003): *Beyond Therapy: Biotechnology and the Pursuit of Happiness,* http://bioethics.georgetown.edu/pcbe/reports/beyondtherapy/beyond_therapy_final_webcorrected.pdf, date accessed 20 August 2012.
Republic of Kenya (2010): National Cancer Control Strategy. 2011–2016. Nairobi: Ministry of Public Health and Sanitation and Ministry of Medical Services.
Romano, G., Bianco, F. and Ciorra, G. (2006): Total anorectal reconstruction with an artificial bowel sphincter. In: Delaini, G. G. (ed.) *Rectal Cancer: New Frontiers in Diagnosis, Treatment and Rehabilitation.* Milan: Springer, pp. 178–182.
Rosman, S. (2004): Cancer and stigma: Experience of patients with chemotherapy induced alopecia. In: *Patient Education and Counseling,* 52(3), pp. 333–339.

Sansom, C. and Mutuma, G. (2002): Kenya faces cancer challenge. In: *Lancet Oncology*, 3(8), pp. 456–458.
Scheper-Hughes, N. and Lock, M. M. (1987): The mindful body: A prolegomenon to the future in medical anthropology. In: *Medical Anthropology Quarterly*, 1(1), pp. 6–41.
Sontag, S. (2001): *Illness as Metaphor and AIDS and Its Metaphors*. London: Vintage.
Stoller, P. (2004): *Stranger in the Village of the Sick. A Memoir of Cancer, Sorcery, and Healing*. Boston: Beacon Press.
Synnott, A. (1993): *The Body Social: Symbolism, Self and Society*. London: Routledge.
Van Dongen, E. (2008): Keeping the feet of the gods and the saints warm: Mundane pragmatics in times of suffering and uncertainty. In: *Anthropology & Medicine*, 15(3), pp. 263–269.
Warren, N. and Manderson, L. (2008): Constructing hope: Dis/continuity and the narrative construction of recovery in the rehabilitation unit. In: *Journal of Contemporary Ethnography*, 37, pp. 180–201.
Watson, P. G. (1992): The optimal functioning plan: A key element in cancer rehabilitation. In: *Cancer Nursing*, 15, pp. 254–263.
Williamson, G. M., Shultz, R., Bridges, M. W. and Behan, A. M. (1994): Social and psychological factors in adjustment to limb amputation. In: *Journal of Social Behaviour and Personality*, 9, pp. 249–268.

# 7
# Singing Better by Sacrificing Sex

*Anna G. Piotrowska*

The biological, cultural, and social phenomenon of singing castrati in European culture of the 17th, 18th, and 19th centuries (Żórawska-Witkowska 2005, pp. 75–109) continues to pose difficult questions. Not only do castrati escape the definitions of gender by mingling male and female codes, they also serve as a starting point for questioning how gender has been (and is) constructed in opera, music, and broader culture. The case of castrati also poses moral questions related to enhancement: what are the ethical limits to interfering with the human body in order to create a more perfect being?

Castrating males in order to generate artistic voices involves sacrificing their sexuality on the altar of music. This raises questions of how to distinguish between 'improving' and 'damaging' the body, and the resulting health condition remains outside of the definitions of 'normality' or 'disability'. We can also ask how such a right to interfere, change, improve, or damage the body of a person for a given reason can be socially established and legitimized. In the historical case of castrati, the ambiguous reaction of society towards them provides no conclusive answer. During the peak of their popularity in the 18th century, castrati were treated as freaks, cripples, and a threat to stable social roles. They were accused of transvestism and homosexuality, and were also ridiculed and laughed at. But on the other hand, they were widely adored, cherished, loved, and pampered.

Castration today may raise a knowing smile or stir up unpleasant feelings, but it nevertheless continues to fascinate and intrigue. In the last decade of the 20th century, largely due to the success of the film *Farinelli* (1994) by Gérard Corbiau, the general perception of a castrato was transformed into that of a 'very pale, brooding romantic figure, with a grown man's speaking voice and a falsetto singing voice' (Harris 1997, p. 182).

Thus constructed and presented to society at large, the figure of the castrato represents, especially in popular culture, a supernatural manifestation of a widely held erotic ideal, with the romantic appeal of a mysterious and mesmerizing hero.

Real – rather than imagined – contemporary castrati (such as *hijras* in South Asia) are still treated with distaste, as 'an embarrassment' (Rosselli 1988, p. 14). If a singing castrato today tried to launch a career in opera, he would probably be treated with curiosity, and the audience 'might take an interest in him as in a freak, but should certainly consider him out of place in any dignified musical environment' (Rogers 1919, p. 413). These strange beings – neither men nor women, neither angels nor monsters, but simply people (musicians, singers, brothers, uncles) – are still waiting for an understanding of their fate.

The rich literature on castrati that appeared in the 20th century focused on their role in the history of modern music, especially opera. However, critical reading of these publications reveals the hidden ambiguity towards the medical condition of castrati, treated by the authors as either a necessary prerequisite or as an explanation for their extraordinary singing abilities. I attempt to answer the question of how castrati have been perceived and presented in 20th century academic writing in light of the fact that their bodies were artificially altered. Did these authors see castration of would-be singers as an act of disablement (in certain spheres of life) or of enhancement (in others)?

## Historical background

### Castration throughout history

The tradition of human castration, although ancient and well-established, remains clouded in a haze of mystery. The intimate and private character of depriving an individual of his external male genitals – known also as gelding or neutering – has always made this topic either shameful or too drastic to be openly discussed in European, male-dominated society. The entirely obscure and legendary origins of castration are attributed to Eastern culture, in particular to the Assyrian Queen Semiramis. By the 4th century AD Roman historian Ammianus Marcellinus had already claimed that she was the first person to castrate boys of tender age (Marcellinus AD 353/1862; Scholz 2001, p. 70). In many ancient cultures, for example Chinese, castration was used in warfare as a means of proclaiming victory (Abbott 1999, p. 318; Mitamura 1970, p. 54). This continued into the medieval age in Europe, when castration was also used as a punishment for seducing women (as in the well-known case of the scholar and philosopher Pierre Abélard, pp. 1079–1142). By contrast, in the Roman Empire, castrati were appreciated and hailed as lovers by promiscuous and rich citizens (Barbier 1996, p. 6).

Castration was often connected with religious aspects of life: from around the 6th century BCE, Anatolian men engaged in self-mutilation in order to avoid the temptations of sin and thus dedicate themselves totally to the service of the mother of the gods, Cybele. In Asia, self-castration would usually take place during the equinox festival, which is regarded as the beginning of spring and is naturally linked with fertility. Christianity, on the other hand, inherited from ancient Greece a rather negative attitude towards castration. The much-celebrated sexual abstinence of Christians was supposed to be the outcome of a strong will, rather than facilitated by the mechanical removal of the testes (which were seen as the source of temptation). However, some Gnostic sects, including the influential Valentinian movement – founded in the 2nd century AD and spreading around the Mediterranean Sea area – believed castration to be the remedy for sin. The First Council of Nicaea in the year 325 finally regulated this issue. It not only prohibited self-castration but also forbade castrati from exercising ecclesiastical functions, allowing admittance only to those who became castrated involuntarily as a consequence of leprosy, epilepsy, gout, or hernia (Milner 1973, p. 251).

Although not very popular within Christianity, castration was often performed by the so-called Skoptsy – a secret sect active within the Orthodox Church in imperial Russia, whose existence came to light in 1772. Their charismatic leader Kondratii Selivanov interpreted the religious imperative for sexual chastity – grounded, he believed, in Biblical passages – in literal terms, as implying the act of physical castration of men and mastectomy for women (Engelstein 1997, p. 2).

Finally, castration as a means to obtain a higher singing voice boomed as a practice in Europe during the Baroque era, for a number of cultural, religious, social, and economic reasons. Castrati became so popular by the early 18th century[1] that their high public visibility turned them into symbols of 'Italian opera as a whole' (McGeary and Cervantes 1999, p. 289). The religious influence of the Roman Catholic Church had a particular effect in Italy, where the words of St Paul – in the First Epistle to the Corinthians, Chapter XVI, verse 34: 'Let women be silent in the assemblies, for it is not permitted to them to speak' – were interpreted literally. As women were banned from singing in church choirs, the demand for higher voices in males emerged. The first choice, training young boys, turned out to be a short-lived solution, since boys were not only difficult to tame let alone teach, but furthermore their voices did not last long, and those with breaking voices had to be replaced as 'often as every three months' (Sawkins 1987, p. 314). The other option was to coach adult men to sing exceptionally high, but falsettists still did not meet expectations as their voices were usually lower than those of sopranos, and had a 'peculiar and unpleasant quality' (Heriot 1956, p. 10), sounding strident and forced. Furthermore, their voices wore out quickly as a consequence of the strain of singing (Giles 1982, p. 75). It is often underlined in the 20th century literature that contemporary with the Italian quest for men with higher singing registers, falsettists were also especially popular in Muslim-dominated Spain (Heriot 1956, pp. 11–12), where they had been introduced in Mozarabic times (i.e., between the 5th and 15th centuries). It is now also believed that many of them were in fact castrati (Heriot 1956, pp. 10–12), and thus the invention of castration for the purpose of singing in religious services can be associated with Spaniards rather than Italians.

Many authors stress the fact that the flourishing of the a cappella style in 15th century church music entailed a parallel demand for a wide range of voices, and the lack of women singers facilitated the infiltration of falsettists and castrati into church choirs, even in the Papal State. Although castrati might have sung there as early as 1562 (Rosselli 1995, p. 34), the first Italian castrati (Barbier 1996, p. 9) were officially admitted to the papal choir in 1599 (Milner 1973, p. 250).

Women, not only in Italy but in the whole of Europe, were generally not only banned from choirs but also discouraged from the stage throughout the 18th century, for fear of their names being linked with prostitution and licentiousness (Heriot 1956, pp. 17, 23–24). This created a demand for higher male voices in the lay sphere as well. Castrati were widely recognized by that time – it was known, for example, that in Constantinople eunuchs were customarily employed as singers. However, there is evidence that even as early as around 1137, a castrato named Manuel already sang in Smolensk (Heriot 1956, p. 10).

## Producing castrati in Italy

Around 1620, the economic crisis in Italy and deindustrialization, rapidly followed by war and the plagues of 1630 and 1656, left the country in a very poor condition (Rosselli 1995, p. 35). Many Italians of comparatively low income, trying to secure a better future for their sons, agreed to have them castrated, foreseeing secure future careers as singers in monasteries. Although castrati usually came from impoverished backgrounds, they did also include the sons of noblemen, minor officials, tradesmen, shoemakers, and musicians.

Parents deciding to sacrifice their sons for a singing career involving castration would usually give (or rather sell) the boy to a teacher, who had him castrated and trained at his own expense. Alternatively, parents would receive money from the musical institution that took care of the boy's future. In rare cases, the parents managed to save enough money to have the operation performed at their own expense, thus assuring themselves of the right to train the boy and later to benefit from the possible fortune he might earn (Heriot 1956, p. 38). The system of recruiting boys for castration was well-developed, and a network of agents worked throughout the whole territory of Italy, affiliated with various educational institutions where boys were taught (Barbier 1996, p. 31). Even though conditions in these places were usually spartan, the castrato boys were considered delicate and therefore enjoyed a better life than other pupils. For instance, they were given warmer rooms to sleep in for fear of chills, they wore nicer clothes, and were better fed, because their health was subject to careful consideration (Heriot 1956, p. 45). The castrato boys were well looked after because investment in their education involved a great deal of money, and though they were subjected to long and strenuous training from early childhood, they did often succeed later as performers or composers (Trinchieri Camiz 1988, p. 182).

Castrato voices were achieved in two ways: either 'naturally' as a consequence of a disease (such as mumps) or an accident, or 'artificially' by deliberate surgical castration (Giles 1982, p. 75). In order to have the castration performed on a boy, his verbal consent was necessary (Barbier 1996, p. 24). However, children undergoing the surgery, who were aged between 8 and 12, arguably would not have been able to appreciate the consequences of the intervention they were asked to consent to. The list of reasons officially given for castration, ridiculous as they might seem, included malformations from birth, mischievous malady, and accidents such as a fall from a horse, a bite from a wild boar or swan, or the kick of an animal or playground friend (Barbier 1996, p. 25; Rosselli 1995, p. 39).

Although in theory castration was illegal, the operation was tolerated in practice and even 'winked at' (Rogers 1919, p. 420). It was performed either (more officially) in hospitals or secretly in dispensaries. The practice was immensely popular in some regions of Italy, including the cities of Naples, Milan, Venice, Bologna, Florence, Lucca, Norcia, Rome, and Lecce (Barbier 1996, p. 27). The surgery itself was usually done with little or no anesthesia (Giles 1982, p. 75), using methods such as drinks containing opium, compression of the carotid arteries, and/or icy baths (Barbier 1996, p. 11). Alternatively, in order to reduce the pain, boys might be soaked in scalding water (Bergeron 1996, p. 170). The surgical procedure, known as orchiectomy, involved making an incision above the pubic bone, through which the testicles could be pulled out through the inguinal canal. The spermatic cord was clamped in two places and then severed in between. In many cases the testes were removed completely (Celletti 1991, p. 109), but sometimes they were left to wither by use of pressure or maceration (Rosselli 1995, p. 36). In rare cases, the penis was amputated as well. Due to the legal emphasis on hygiene, mortality was not high: authors claim that there were few cases of death caused by castration as the surgery was a 'routine one' (Rosselli 1995, p. 37). However, other sources speak of a fatality rate ranging from 10 percent to as high as 80 percent (Barbier 1996, p. 11).

The condition of a castrato, described as primary hypogonadism, involved changes that affected both body and psyche (André 2006, p. 28). The most sought-after feature of castration was, of course, connected with sustaining the young boy's high singing voice. The castrato's vocal cords remained the same size as a boy's (André 2006, p. 30), as the operation prevented the growth of the larynx, which did not descend, leaving the vocal cords 'closer to the cavities of resonance' (Barbier 1996, p. 16). The voice was thus prevented from breaking, and there was no development of an Adam's apple. Castrati's ribcages were, however, well developed, leaving ample room for large lungs, useful for singing. The amount of air they could take in allowed them to sing long phrases on one breath, giving spectacular effects. Specialized castrati were said to be capable of sustaining a note for a whole minute without taking breath (Barbier 1996, p. 95). It was also thought that the voice of a castrato joined 'the sweetness of the flute and the animated suavity of the human larynx' (Barbier 1996, p. 36), and was characterized as peculiar, certainly different from a woman's voice (Heriot 1956, p. 14).

In terms of other physical effects from castration, the reported long lives of many castrati led to the belief that one of the consequences of emasculation was the prolongation of life expectancy.[2] The manifestation of secondary sexual features was also disturbed in the bodies of castrati; they never grew beards or moustaches and lacked body hair (with the possible exception of the pubic areas). They did, however, have a tendency to develop unusually long limbs, and were often either extremely tall and thin or obese. Facilitated by the activity of female hormones in the body that were not balanced by the presence of testosterone, castrati tended to deposit fat on their hips, neck, thighs, and chest, which might have led to their slightly feminine look. With regards to the psychological consequences of castration, the temperament of castrati was described as difficult: many apparently suffered from frequent and uncontrolled fits of anger and were believed to suffer from neurasthenia and melancholy, which deepened with age (Barbier 1996, pp. 149, 209).

## Fame on and off the stage

Actors performing in opera houses were recruited from among professional singers in the church choirs (Rogers 1919, p. 414), where most young castrato boys performed until their patrons or teachers decided that they could appear on stage. Young boys debuting on stage would usually adopt pseudonyms. It was a widespread custom for castrati to decide upon their stage names while still in training, often choosing diminutive forms of the name of their teacher as a token of gratitude for the master's influence over the boy's life (Barbier 1996, pp. 83–84).

The more talented boys made their public debuts between the ages of 12 and 16 years of age (Barbier 1996, p. 86). These young castrati – with their freshness and ambiguity of looks – were cast in female roles. However, they frequently longed to be cast as great heroes such as Alexander the Great or Orpheus, and when they grew older they often did perform heroic male roles (Barbier 1996, pp. 87–88; Heriot 1956, p. 19). Half-human, half-godlike, they seemed predestined to sing these parts. On his successful debut, a new castrato could count on quickly forming a group of devoted followers, both male and female. Fascinated with his musical talent and singing skills, admirers wrote poems about their favorite castrato, sent him flowers or more valuable presents, and perhaps even followed him around. These groups of what could be called fans were sexually attracted to the castrato himself, discussed his voice, attended all his performances, and worshipped the newborn star. As a symbol of the castrato's new social position, even prominence, portraits were painted (Trinchieri Camiz 1988, p. 183). Various cliques of avid supporters were often antagonistic towards one another, with heated discussions and arguments taking place between fans of different castrati. Surprisingly, although many excelled, not all castrati were that good at singing or acting; but the fans and publicity these lesser performers enjoyed still enabled them to maintain a high position among other singers.

Winning the hearts of a local public could lead to international fame. Once established on the stage, many – though not all – castrati carefully built up their image, which consequently led to the stereotype of the castrato as a vain and extravagant opera star. Castrati jealously guarded their status as stars. There are accounts of feuds between castrati, intrigues, and other forms of open or hidden animosity. Castrati blackmailed their patrons, refusing to perform in the presence of other castrati, or even fought with each other in public (Barbier 1996, p. 149). More refined methods of ridiculing opponents included targeting sarcastic remarks at them. The star behavior of some castrati – described as extravagant and exhibitionistic – extended to their private lives as well. As in any social group, some castrati were vulgar, quarrelsome, or exhibitionist. Yet the image of castrati as sexually rampant, 'colossally vain, extravagant and temperamental', was created primarily for the publicity needs of only those castrati who cherished international fame (Rosselli 1988, p. 173). It seems, however, that the majority of castrati, singing for example in choirs, who are rarely discussed by present-day historians in these terms, did not have such an extravagant temperament, and rumors about their outrageous behavior and irresistible sexual allure referred only to a few.

The extravagance of opera and the 'sexual license surrounding the castrati' were reported by their contemporaries with similar attitudes of disapproval and attraction (Abbott 1999, p. 334; Harris 1997, p. 185). Some castrati were renowned for their numerous – real or attributed – affairs with aristocratic ladies and treated as sex symbols; they were mostly appreciated by women from 'society', who happily took them as lovers. Castrati were said to arouse sexual feelings to such a degree that their most ardent admirers followed them wherever they performed (Rosselli 1995, p. 52). Amorous intrigues involving castrati and their well-born admirers were rumored, often in exaggerated form, adding to their star status (Giles 1982, p. 77). Homosexual adventures were also attributed to castrati, but while their romantic adventures with women were widely reported – even spotlighted – alleged homosexual relationships were kept secret (Barbier 1996, p. 151). Young castrati's sexual appeal is sometimes explained by researchers with reference to the fact that 'in the early modern period [...]among men whose tastes included homosexual sodomy, boys were the generally preferred partners' (Freitas 2003, p. 210). It is assumed by them that the figure of a castrato 'represented a theatrical imitation of this erotically charged boy' (Freitas 2003, p. 214).

In spite of the fact that they were surrounded by so many people, it was loneliness that posed the 'chief hazard' in most castrati's lives (Rosselli 1995, p. 53). Misunderstood by admirers, treated as a source of financial security by relatives, and unable to raise their own families, castrati frequently developed a bittersweet attitude towards life and its pleasures. This melancholic state often deepened towards the end of their lives, when the peak of fame was over and all that they were left with were memories, or a small fortune at their disposal. Furthermore, despite earning a lot of money, being in the public eye, and often enjoying a promiscuous life, castrati were sometimes presented as a symbol of the degradation of Italy (Heriot 1956, p. 53) and were subject to criticism from their contemporaries. Castration itself was also the subject of mockery (both verbal, in pamphlets, and visual, in caricatures). Castrati were ridiculed in many circulating satires, using animal related synonyms such as *cappone* (capon) or *castrone* (gelding). Castrati were also pitied since it was felt that their 'humanity' had been violated, as the reproductive potential of a castrato was removed and their sterility became their defining feature (Lambert 1986, pp. 167–168).

## An attempt to define the phenomenon of castrati in Italian opera

### Gender qualities of castrati

Young boys learn their gender roles during the tender age of adolescence, and many child development theories suggest that the development of masculinity in males (and femininity in females) is of primary importance during this time. However, in response to the demands that society placed on them as individuals preparing to adopt the profession of singers (Lerner et al. 1981, pp. 709–722), the gender model that young castrati might have adopted 'extend[ed] beyond masculine and feminine sex typing to androgyny and sex-role transcendence' (Lamke 1982, p. 1530). The gender qualities represented by castrati in the popular arena of Italian opera are considered below in terms of the specific attributes of voice, cross-dressing, and sexual behavior.

1. Voice

The voices of castrati, who were called 'Italian sopranos', were considered their female element, and were perceived as highly sensual (Barbier 1996, p. 192). Treating castrati as women stemmed from a tradition of associating an individual's vocal register with his or her sex. The range of the voice was understood as an indisputable attribute of sex, given by nature and thus its permanent and unmistakable feature (Koestenbaum 1991, pp. 205–234). In the history of European culture, women certified their gender and asserted their femininity by performing high-pitched songs. The high female voice became treated as a sexual trait, more powerful even than appearance; in Greek mythology, for instance, sirens seduced sailors, including Odysseus returning to Ithaca, with their enchanting voices. The polarization of male and female voices was blurred, however, in the age of castrati. The broad range of their voices (as well as the voices of those few women who performed in the early times of opera, usually specializing in male roles) encompassing over three octaves, were theorized as 'belonging to neither male nor female' (Wood 2000, p. 86), thus they introduced a new quality, 'a synthesis, not a split' (Wood 2000, p. 86). Elizabeth Wood suggests calling these border-crossing voices 'Sapphonic' and categorizes them as a 'transvestic enigma' (Wood 2000, p. 86).

2. Cross-dressing

The distortion of visual codes signifying masculinity and femininity were perhaps best observed in cross-dressing practices. Castrati wore female attire only occasionally, and in the majority of iconographic images of them they are represented wearing men's clothes (André 2006, p. 28). Yet sometimes, female stage roles performed by castrati were continued outside the theatre, and the costume became part of their identity. Furthermore, the gestures, mimicry, and movements of some castrati were so feminine that the overall image was more that of a woman than a man.[3] Features such as cross-dressing and taking on some female characteristics were often treated as examples of transvestism (Barbier 1996, p. 152).

The provocative apparel that many castrati wore, however, was designed not to heighten their sexual identity, nor to hide it, but in fact to blur it.[4] Due to their androgynous looks, castrati could pass either as women or as men. Even though most people knew clearly that they were not women, the thrill and surrealism of the situation was still achieved via the deception, building up an erotic atmosphere. Cross-dressing by castrati unraveled contemporary sexual signifiers of ambivalence, androgyny, and eroticism (Cressy 1996, p. 439).

3. Sexual behavior

The sensuality of castrati's singing skills and acting, coupled with the reputed extravagance of their personae, made castrati desired by all in society. They especially enjoyed a reputation as ladykillers, with supposedly only French women treating them as impaired and wondering why anyone would prefer 'the half rather than the whole' (Barbier 1996, p. 193). The hybrid sensuality of castrati was evident in their looks combined with their voice, as well as their exoticism as creatures different from all others. Although onstage castrati were ascribed either male or female roles in reference to their sexuality, they were also perceived – as modern texts suggest – as people who combined female and male features in one body (Barbier 1996, p. 17). In a capricious society that sought novelty, the passionate castrato constituted the perfect symbol of a sexual creature (what is nowadays called a 'sex symbol'). Having an affair with a castrato meant an adventure and yet it posed no problematic consequences in the form of unwanted pregnancy. The promiscuity of castrati – a feature attributed to them by their contemporaries – constituted one of the traits associated with their star status, as mentioned above. Hence they were objects of the fantasies and desires of aristocratic admirers, often wealthy ladies for whom opera and singers were a part of their lives.

The immensely erotic and provocative presence of castrati, as opera stars, enabled them to be simultaneously attractive to both men and women in the audience, though instances of homosexual relations were not openly advertised. Castrati were, however, also accused of low morals and condemned for attempts at various sexual activities. They were objects of both pleasure and shame, and their sexual appetite was viewed as inappropriate because it could never lead to conceiving a child; since castrati could not father children they were often described by contemporaries as 'dry trees' (Barbier 1996, p. 167). Hence, their sexual life was seen within the category of lust and fulfilling mundane desires; castrati were associated with debauchery, voluptuousness, and lechery and were thought to be denied love and normal human relations. Even the prospect of a castrato marrying a woman was treated with suspicion, and the Roman Catholic Church strongly opposed it.

As castration is a surgery on the reproductive system, it of course affects the sexual life of castrati. The atrophied testes meant that castrati could not procreate, as they no longer secreted spermatozoa. However, since the production of seminal fluid was not disturbed, a castrato could undertake sexual activities and perform coitus (Celletti 1991, p. 109). The 19th century French surgeon Benedetto Mojon, in a short treatise on castration, asserted that 'when the operation has been performed closer to the age of puberty, his equipment – or what's left of it – will more nearly resemble that of a normal man, with one notable exception: erection takes place much more frequently than in the case of non-castrated men'. Nevertheless, the absence of testosterone could affect the sexual act due to a weakened erection (Celletti 1991, p. 109). The intensity of the experience as well as the regularity of sexual activity were obviously drastically altered compared to non-castrated men. Their sexual capacities were reduced yet not destroyed as a result of the operation, however the reported level of sexual desire among castrati varied beyond any meaningful generalization. The question of whether castrati could form romantic attachments to women, and consequently marry them, has been a recurrent topic in the 20th century literature. In theological thought, castrati – as unable to produce offspring – were not allowed to marry (Sherr 1980, p. 56). There are, nevertheless, a few reported examples of marriages between castrati and women.

However, in direct contrast with the 'common' stereotypes of castrati being sexually promiscuous and extravagant opera divas were those numerous castrati whose assumed celibacy was linked to religious music. In fact many castrati did become monks and sang in church choirs (Barbier 1996, p. 129). In the 17th and even the 18th centuries, when Italy was struggling with its poor economy, the choice of an ecclesiastic life for a son seemed obvious. Castrati who became monks, and thus made a vow of celibacy, found themselves in a privileged position as they did not have to resist bodily temptations; after all, as late as the 18th century, monks themselves were popularly viewed as castrati who had not been 'operated upon' (Rosselli 1995, p. 35) and thus had to exert a greater force of will in order to be celibate. These monk castrati were celibate, and in times of considerable poverty sexual abstinence was considered a birth control measure (Rosselli 1995, p. 36).

## Union of sexes

A critical reading of 20th century writings on the phenomenon of castrati reveals the tendency to describe them as representing a union of sexes – either as man *and* woman, or alternatively as an amalgamation of male, female, and childlike qualities. Some alternatively present castrati as representing an androgynous form of male embodiment.

### 1. Hybridity

The castrati's looks, as well as their voices and the knowledge of their (dramatic) origin, meant that they were often defined with reference to their gender. The span of castrato popularity in Italy and elsewhere in Europe – from the early 17th to the end of the 19th century – parallels the overall changes in attitudes towards sex itself. Thomas Laqueur suggests that in Europe in pre-modern times – that is, from the Middle Ages up until the 18th century – the predominant understanding of sexual difference saw the two sexes as one, one form of 'elastic sex', where male and female were inversions of each other (Laqueur 1992, pp. 124–125). The single-sex model's explanation for the visual differences between male and female were that the visible outer genitalia of a man were hidden inside the female. Masculinity and femininity were like the opposite ends of the same line and only the concentration of specific features decided which label was attached to an individual: as a (perfect) male or as a (less perfect) female. Differences between men and women were complementary, as they were all parts of the same biological whole. In light of Laqueur's theory, castrati were the embodiment of the union of the two sexes (André 2006, p. 46; Freitas 2003, p. 204). Their attire, voice, manners, etcetera, constituted the blend of male and female without necessitating the labeling of an individual as either one or the other. Castrati were believed to combine 'masculine presence and feminine grace' (Barbier 1996, p. 107).

In 'modern' times – starting in the 19th century – according to Laqueur, a new paradigm for the sexes was created. Women ceased to be seen as the imperfect reflection of men but started to be perceived as their complete opposite. Soon it was widely accepted that men and women represented opposite sexes, and the two-sex model was introduced. This spawned a new perception of castrati. Their sex became perceived either as a strange hybridity or as a lack of sex altogether: castrati were representatives of a 'sexless tribe' (Rogers 1919, p. 413).

However, the most recent studies into the history of castrati tend to acknowledge the existence of a 'third' gender, as represented by castrati. Naomi André, in her book *Voicing Gender*, suggests that castrati combined masculine and feminine codes, with the interaction between these two leading to the creation of a new category – a third zone, where a + b does not equal their sum ab, but constitutes c (André 2006, p. 48). The third zone is not merely the synthesis of the two sexes but introduces new elements. This third gender thus encompasses neither masculine nor feminine exclusively, but opens up a space of possibility (Garber 1992, p. 11). The biology of sexual reality and sexual ideologies (for instance, belief in the dualism of feminine and masculine) are confronted with socially determined gender identity, whereby cultural distinctions have been 'naturalized' (Lambert 1986, p. 162). So-called third gender people are physically either men or women, yet they transgress the cultural boundaries of the behavioral codes attributed to a given sex. Examples of the idea of a third gender can be found today as well as in the past (Herdt 1996): in literature (Virginia Woolf's *Orlando*, to name but one), film (the 1995 film *To Wong Foo, Thanks for Everything! Julie Newmar*, directed by Beeban Kidron), the visual arts (see the sculptures of the Borghese Hermaphroditus), and music (e.g., Pete Burns, the leader of the British band Dead or Alive).

The definition of a castrato as a union of man and woman is sometimes enlarged to include the child (Freitas 2003, p. 214). The childlike features attached to castrati refer predominantly to the angelic quality of their voices; the voice of a boy is sometimes seen as being trapped within the adult body of a castrato. Some modern authors have therefore suggested that a castrato in fact 'embodied the trinity – man, woman and child' (Barbier 1996, p. 17), while contemporaries sometimes characterized castrati as 'effeminate children' (Barbier 1996, p. 181). The inclusion in their persona of childlike features is also associated by some with a child's purity and virginity, such that the death certificates of some castrati described them as 'virgins' (Barbier 1996, p. 222).

2. Juxtaposition of sexes in the figure of the castrato: Carnivalization
The juxtaposition of opposites is a characteristic feature of local Italian customs in the period of carnival, but can also, in my opinion, be seen as part of the European Baroque era more generally, during which – especially in opera – castrato singers were favored, performing not only male roles but female roles as well. Women on operatic stages were cast in male roles, or alternatively as castrati pretending to be women. The masquerade was part and parcel of the operatic world, in which carnival seemed to last forever. Opera as a form of entertainment provided a sphere for the juxtaposition of social roles, and constituted a space for fulfilling various desires, be they sexual, economic, or social. The charade, in which women pretended to be men, and castrati pretended to be women – and were treated as kings despite their poor origin – enabled an aristocratic audience to detach themselves from everyday life.

3. The androgyny of castrati
Some authors argue – based on the significant embodiment of both feminine and masculine characteristics of castrati – that the castration of boys before puberty resulted in an androgynous form of male embodiment (Trinchieri Camiz 1988, p. 174). It can be claimed that in their conduct of behavior 'masculinity and femininity are complementary, not opposite positive domains of traits and behaviors' (Powell and Butterfield 1979, p. 396).

### Unity of opposites

The literature offers yet more explanations of castrati beyond the tradition of viewing them as a combination of sexes or as hybrid creations. This interpretation has two variants:

1. Angel and monster
The unearthly voices of castrati meant that they are seen as the link between God and humankind. Castrati defied 'all the laws of morality and reason to achieve the impossible union of monster and angel' (Barbier 1996, pp. 17, 242). A typical castrato – growing to unusual height, with a body sometimes severely out of proportion – was at the same time often endowed with beautiful, angelic features due to the fact that subcutaneous fat often localized on the face, preserving a childish appearance. Thus castrati physically resembled an awkward combination of monstrous body and angelic face.

2. Visual and aural
From its beginnings, the hybrid nature of musical aesthetics in opera combined male and female, sound and sight, as well as ancient elements – resurrected from Greek drama – and modern, exotic, and more familiar ones. For some observers, castrati represent the 'close connection between sound and sight' (André 2006, p. 23), as in the body of a castrato, the aural element found its ultimate visual representation. In the Baroque era when both aural and visual aspects of music performance were not static, but constantly being reinterpreted, castrati helped to redefine them anew.

## Social reactions and attitudes towards castrati

Castrati, in escaping unanimous definition and categorization, aroused curiosity and interest among their contemporaries, as well as disrespect and contempt. Modern writings describe these reactions openly; the authors, however, often restrict themselves to merely citing facts and avoid classifying social attitudes towards castrati. Nevertheless, the general reactions towards castrati, as reported by modern historians, can be divided into the positive – including admiration, and revealing some undertones of fascination with the phenomenon – and the negative – concentrating on ambiguous emotions stirred up by the figure of the castrato.

### Positive

1. The superiority of higher voices
The high voices of castrati were perceived as being superior to those possessed by other males in the light of the 'higher-is-better' theory (Freitas 2003, p. 198). The preference for higher voices meant that the higher a castrato could sing, the more he was paid. In the operatic world, the preference for higher voices continued throughout the 19th century,

when sopranos were still considered better than altos, and tenors were valued more highly than basses (Rosselli 1995, pp. 34–35).

2. Supremacy over women

As cruel as castration may seem to us, it was nevertheless considered acceptable by contemporaries as a means to produce high-pitched singers, from the beginnings of opera as a genre in the late 16th century until about 1800. Furthermore, castrati were hailed as superb musicians; even the word castrato was initially understood by Italians as a synonym for a musician (Giles 1982, p. 74). Furthermore, castrati were seen as being superior to women singers, for at least two reasons. The first was the voice itself – with a broader range and power, castrati were capable of performing with a strength and breath control unknown to women singers. The high voice of a castrato was also valued for its greater endurance and praised for its agility and ability to thrill (Trinchieri Camiz 1988, p. 173). The second reason was that they could easily perform female roles at a time when women were banned from the stage. Some contemporary intellectuals, including Johann Wolfgang Goethe, preferred castrati over women on the stage. Goethe advocated for their presence by claiming that art is an imitation of nature – similar to the embodiment of the castrato (Heriot 1956, p. 26).

### Negative

1. The castrato as a symbol of fear

Castrati were not always revered or celebrated, and sometimes they were ridiculed by both men and women: by women for being only half rather than a whole man; by men perhaps out of the fear of being castrated themselves. Some authors suggest that men are afraid that for reasons over which they would have no control, 'they may be castrated, that they may be laughed at, that accepting female roles in the theatre, by choosing to wear female attire', and so on, castrati symbolized 'subversion, as well as modification, recuperation, and containment of the system of gendered patriarchal domination' (Cressy 1996, pp. 438–439). Hence the rhetoric of anxiety and fear used by their contemporaries while describing castrati refers to the danger of upsetting patriarchal values.

'They may die, and have presumably always been terrified that castration would nullify their dominant social position' (Raitt 1980, p. 418). In psychoanalytical assumptions the fear of castration and of raising another man's children seems a worry that haunts many men (Penuel 2004, p. 267), and it can be argued that in the figure of the castrato it found its ultimate projection. Furthermore, it is claimed by some authors that the inversion of sex roles, even if festive, subverts the hierarchical structure of gender relationships (Schleiner 1988, p. 605). Castrati thus represented a threat to the ideal order in which 'real' – meaning uncastrated – men dominate.

Castrati also symbolized the fear of not belonging. As representatives of neither sex, they belonged to neither gender group. Neutrality in this sense was an abnormal condition, even frightening when observed in hermaphrodites. The underprivileged position of hermaphrodites, however, was associated with their being freaks of nature, and this prevented them from becoming the object of laughter. But the artificially crafted castrati were not allowed such protection. The mockery and disdain targeted at them reveals the social awareness of the degree of damage caused by the surgery in such a delicate sphere as human sex and gender.

2. Castrati as artificial creatures

Castrati are viewed either as a perfect blend of the sexes, or as people with no sex at all. This is described as inhuman (Barbier 1996, p. 166). Artificiality could be also detected in their voices, 'which can hardly be described as natural' (Giles 1982, p. 74), and their possession of singing abilities beyond normal human limits. A castrato, then, is described as a 'singing machine constructed simply and solely by making use of the laws of biology' (Celletti 1991, p. 108).

Castrati are thus the products of their creators – parents, surgeons, and teachers–with no choice over their own lifestyle. They are also products of their time, created in the name of art – as in the myth of Pygmalion – attesting with their own lives to 'the role of art as a means of transcending cultural limitations' (Lambert 1986, p. 170). Their ambiguous sexuality coupled with artistic excellence serves as living proof of the thesis that 'in music, as with sex, virtuosity for its own sake brings only empty pleasure, but music and sex with true feeling are transcendent experiences' (Harris 1997, p. 180).

## Castrati: Escaping definition

Castrati escape categorization or simple definition since they represent the 'confusion of biological sexual identity with socially determined gender behavior that in turn depends on the rigidly defined opposition male/female' (Lambert 1986, p. 166). Mainstream society, which tended to label people as men or women, or as people with a vagina or a penis, became confused. Such traits as a castrato's high voice or acts of cross-dressing acquired their social value depending upon their sexual association with either men or women; yet castrati's hybrid sexuality prevented onlookers from easily classifying their behavior and appearance, and thus prevented their being classified as persons.

Castration will never cease to both frighten and attract, as it deeply interferes with ideas of human identity and embodiment. The historical case of castrato singers raises many questions: who was entitled to perform this surgery for the sake of art, and what should be the limits of human transformation for such reasons? Perhaps, however, these questions may be irrelevant; castrati might represent the longing for an ideal, which is to be simultaneously a man and a woman, while also holding on to a child's traits – eternal youth. To achieve this dream, people were prepared to go to extremes and demand a high price from those they fashioned to embody this ideal.

It has not been my goal to provide a straightforward answer to the question of whether a singing castrato represents a case of disablement or enhancement. From my critical reading of modern texts on castrati comes the observation that for many authors this differentiation seems irrelevant, or at least of lesser importance, compared to the effect gained by castrating singers, that is obtaining extraordinary singing abilities. External intervention into the bodies of would-be castrati is mentioned without analysis or classification of the categories of improvement or degradation. The historical phenomenon of castrati and its interpretation in the 20th century show how close notions of disablement and enhancement are in terms of human bodies, and prove that in certain circumstances they may in fact be one and the same.

## Notes

1. Apparently the only exception was in the Paris Opera, where castrati never sang, although six castrati were employed at the French court (Sawkins 1987, p. 319).
2. For example, castrato Antonio Bagniera, doyen of the court castrati in the late 17th and early 18th centuries, survived to the age of 102 (as discussed in Sawkins 1987, p. 321).
3. Giuseppe Casanova, giving an account of his first meeting with a castrato, described him as a beautiful woman, whom he thought an adorable creature, before being told of his true identity as a castrato. He wrote that 'at the appearance of his hips I took him for a girl in disguise' (see Giles 1982, p. 78).
4. In the 1830 novel by Honoré de Balzac, entitled *Sarrasine*, the castrato La Zambinella not only plays female roles in the opera, but also dresses and

presents himself as a woman, causing the Parisian sculptor Sarrasine to fall desperately in love with him. For the artist, La Zambinella embodies the perfect femininity (see Barthes 1974).

## References

Abbott, E. (1999): *A History of Celibacy.* New York: Scribner Press.

André, N. (2006): Voicing Gender: Castrati, Travesti, and the Second Woman *in Early-Nineteenth-Century Italian Opera.* Bloomington and Indianapolis: Indiana University Press.

Barbier, P. (1996): *The World of Castrati. The History of an Extraordinary Operatic Phenomenon.* London: Souvenir Press.

Barthes, R. (1974): *S/Z.* New York: Hill and Wang.

Bergeron, K. (1996): The castrato as history. In: *Cambridge Opera Journal*, 8(2), pp. 167–184.

Celletti, R. (1991): *A History of Bel Canto*, translated by Fuller, F. Oxford: Clarendon Press.

Cressy, D. (1996): Gender trouble and cross-dressing in early modern England. In: *The Journal of British Studies*, 35(4), pp. 438–465.

Engelstein, L. (1997): From heresy to harm: Self-castrators in the civic discourse of late Tsarist Russia. In: Teruyuki, H. and Kimitaka, M. (eds.) Empire and Society: New Approaches to Russian History. Sapporo: Hokkaido University Slavic Research Center.

Freitas, R. (2003): The eroticism of emasculation: Confronting the Baroque body of the castrato. In: *Journal of Musicology*, 20(2), pp. 196–249.

Garber, M. (1992): Vested Interests*: Cross-dressing and Cultural Anxiety.* New York: Routledge.

Giles, P. (1982): *The Counter Tenor.* London: Frederick Muller Limited.

Harris, E. T. (1997): Twentieth-century Farinelli. In: *The Musical Quarterly*, 81(2), pp. 180–189.

Herdt, G. (ed.) (1996): Third Sex, Third Gender: Beyond Sexual Dimorphism in Culture and History. New York: Zone Books.

Heriot, A. (1956): The Castrati in Opera. London: Secker & Warburg.

Koestenbaum, W. (1991): The queen's throat: (Homo)sexuality and the art of singing. In: Fuss, D. (ed.) *Inside/Out: Lesbian Theories, Gay Theories.* New York: Routledge.

Lambert, D. G. (1986): S/Z: Barthes' castration camp and the discourse of polarity. In: *Modern Language Studies*, 16(3), pp. 161–171.

Lamke, L. K. (1982): The impact of sex-role orientation on self-esteem in early adolescence. In: *Child Development*, 53(6), pp. 1530–1535.

Laqueur, T. (1992): *Making Sex: Body and Gender from the Greeks to Freud.* Cambridge: Harvard University Press.

Lerner, R. M., Sorell, G. T. and Brackney, B. E. (1981): Sex differences in self-concept and self-esteem of late adolescents: A time-lag analysis. In: *Sex Roles*, 7, pp. 709–722.

McGeary, T. and Cervantes, X. (1999): Farinelli to Monticelli: An opera satire of 1742 re-examined. In: *The Burlington Magazine*, 141(1154), pp. 287–289.

Milner, A. (1973): The sacred capons. In: *The Musical Times*, 114(1561), pp. 250–252.
Mitamura, T. (1970): *Chinese Eunuchs: The Structure of Intimate Politics*, translated by Pomeroy, C. A. Tokyo: Tuttle Publishing.
Penuel, S. (2004): Castrating the creditor in 'The Merchant of Venice'. In: *Studies in English Literature 1500–1900*, 44(2), pp. 255–275.
Powell, G. N. and Butterfield, D. A. (1979): The 'good manager': Masculine or androgynous. In: *The Academy of Management Journal*, 22(2), pp. 395–403.
Raitt, J. (1980): The 'Vagina Dentata' and the 'Immaculatus Uterus Divini Fontis.' In: *Journal of the American Academy of Religion*, 48(3), pp. 415–431.
Rogers, F. (1919): The male soprano. In: *The Musical Quarterly*, 5(3), pp. 413–425.
Rosselli, J. (1988): The castrati as a professional group and a social phenomenon 1550–1850. In: *Acta Musicologica*, 60(2), pp. 143–179.
Rosselli, J. (1995): *Singers of Italian Opera. The History of a Profession.* Cambridge: Cambridge University Press.
Sawkins, L. (1987): For and against the order of nature: Who sang the soprano. In: *Early Music*, 15(3), pp. 315–324.
Schleiner, W. (1988): Male cross-dressing and transvestism in renaissance romances. In: *The Sixteenth Century Journal*, 19(4), pp. 605–619.
Scholz, P. O. (2001): *Eunuchs and Castrati: A Cultural History.* Princeton: Markus Wiener Publishers.
Sherr, R. (1980): Gugliemo Gonzaga and the castrati. In: *Renaissance Quarterly*, 33(1), pp. 33–56.
Trinchieri Camiz, F. (1988): The castrato singer: From informal to formal portraiture. In: *Artibus et Historiae*, 9(18), pp. 171–186.
Wood, E. (2000): On the Sapphonic voice. In: Scott, D. B. (ed.) *Music, Culture, and Society. A Reader.* New York: Oxford University Press.
Żórawska-Witkowska, A. (2005): Głos utracony-kastrat jako fenomen fizjologiczny, artystyczny, kulturowy. Barok. In: *Historia-Literatura-Sztuka*, 23, pp. 75–109.

# Part III

# Utopian Ideas and Real Embodiment

# 8

# Mood Enhancement and the Authenticity of Experience: Ethical Considerations

*Lisa Forsberg*

> Suppose there were an experience machine that would give you any experience you desired. Superduper neuro-psychologists could stimulate your brain so that you would think and feel you were writing a great novel, or making a friend, or reading an interesting book. All the time you would be floating in a tank, with electrodes attached to your brain. Should you plug into this machine for life, pre-programming your life's desires? [...] Of course, while in the tank you won't know that you're there; you'll think it's all actually happening... Would you plug in? What else can matter to us, other than how our lives feel from the inside?

This section of Robert Nozick's *Anarchy, State and Utopia* (1974, p. 42) introduces the Experience Machine, a famous thought experiment that Nozick used in an attempt to refute mental state theories of well-being. Nozick conceived the Experience Machine to put forward an argument along the following lines: in spite of the fact that we will experience more pleasure if we plug into the Experience Machine than if we do not, we have good reasons not to plug in. This suggests that experiencing as much pleasure as we can is not all that matters to us. According to Nozick, there are three reasons why we should not plug into the Experience Machine. First, we want to *do* certain things, as opposed to merely have the experience of doing them. According to Nozick, '[i]t is only because we first want to do the actions that we want the experiences of doing them' (1974, p. 43). Second, it is important to us to *be* a particular kind of person, rather than merely have the experience of being that person, while in fact being 'an indeterminate blob', which is what those who plug into the Experience Machine are, according to Nozick (Ibid.). Third, if we were to plug into the Experience Machine, we would be limited to a version of reality that is man-made. In the Experience Machine there would not be any '*actual* contact with any deeper reality, although the experience of it would be simulated [original emphasis]' (Ibid.). All of these reasons are, to some extent, concordant with a line of reasoning according to which authentic or real experiences or mental states are more valuable than experiences that are brought about through the use of an experience-inducing tool. Those who oppose the use of mood enhancement (ME) technologies often employ a similar line of reasoning.

In recent years, the prospect of enhancing our cognitive capacities or mood has received widespread attention. It has been suggested that pharmacological means that could be used to alter many aspects of our cognitive capacities or mood already exist. With regard to cognition, it has been suggested that pharmacological agents, such as methylphenidate (used to treat narcolepsy and attention deficit hyperactivity disorder) and modafinil (used to treat narcolepsy and shift-work sleep disorder), could be used as 'cognitive enhancers', because of their concentration and wakefulness promoting properties. With regard to mood, it has been proposed that psychopharmaceuticals such as selective serotonin reuptake inhibitors (SSRIs) – one of the most common class of antidepressants – may positively affect mood even in individuals who do not suffer from mood disorders. For example, in *Listening to Prozac,* Peter Kramer (1993) concluded from his own experiences with patients that psychotropic drugs may have the potential to improve minor mood problems or enhance temperament in individuals who are not clinically depressed. He refers to this as 'cosmetic psychopharmacology', and argues that such interventions could be for mood what cosmetic surgery is for physical appearance. The idea of ME or cosmetic psychopharmacology has since been considered by a large number of commentators. Although the efficacy of such agents in healthy individuals is a matter of debate, let us assume, for the sake of argument, that reasonably efficacious mood enhancers existed. Would their use, as it has been argued by some, risk undermining the authenticity of our experiences?

In the present paper, I consider two arguments against the use of ME. Both of these arguments claim that ME somehow undermines the authenticity of human experience or existence, and that this is somehow objectionable. Both these arguments rely on the assumption that a connection to reality is necessary for an experience to be authentic, and that the lack of such a connection to reality is problematic. According to the first argument, authentic experiences are *qualitatively superior* to inauthentic ones. According to the second argument, authentic or natural emotional responses to life events are useful, and ought therefore not to be interfered with. I will argue that both arguments fail to capture (what may be) compelling reasons for us to restrict the use of (some forms of) ME.

## The qualitative superiority of authentic experiences

The first argument holds that authentic experiences are qualitatively superior to inauthentic experiences. In order to assess the force of this argument, a closer look at the notion of authenticity is needed.

Despite the prevalence of the notion of authenticity in the academic literature concerned with the permissibility or desirability of (different forms of) enhancement, the nature of authenticity remains elusive, its role and importance dubious, and its relation to enhancement a matter of controversy. Indeed, as Erik Parens has noted, '[a]nyone who has used the word "authenticity" or has tried to track how others use it, knows how slippery it is' (2009, p. 41). As Parens has observed, those who oppose enhancement tend to argue that it is objectionable because it undermines the authenticity of those who use it, while proponents of enhancement argue that its very purpose is for individuals to be able to create a life that is more authentically their own.

Authenticity is often understood to mean something along the lines of 'being true to oneself'. But what does this mean in practice, and how do we know whether we are true to ourselves or not? According to Neil Levy, authentic individuals 'do not passively accept social roles imposed upon them. They do not simply select between the conventional ways of living that their society makes available. Instead, they look for and actively create their *own* way, by reference to who they, truly and deeply, are' [original emphasis] (2007, p. 74).

It seems questionable, however, whether we would be able to distinguish authentic lives and individuals from those that are inauthentic. How would we know whether someone had actively created 'their own way of life, by reference to who they, truly and deeply, are'? This also applies to our subjective experiences: how would we ourselves know who we, truly and deeply, are? Levy has, for example, argued that the ideal of authenticity is specific to modern societies:

> Authenticity [...] could not exist in premodern societies, in which social roles were relatively few, and people had little freedom to move between them. [...] Authenticity requires the growth of cities, and

> the consequent decrease in the social surveillance and mutual policing characteristic of village life. In the anonymity of the city, people [are] free to remake themselves. They could, if they wished, break free [...] from the expectations of their family, their church, their friends and even of social conventions, and remake themselves in their own image.
>
> (Ibid., p. 74)

This account of what it means to be an authentic individual appears to be very demanding, since it requires individuals to break free from societal norms and patterns of life. First, this account seems to presuppose a very low degree of determinism in the world. Second, it may seem somewhat unrealistic to assume that individuals would be able to 'break free [...] from the expectations of their family, their church, their friends and even of social conventions, and remake themselves in their own image'. Indeed, one might question the extent to which many of us lead lives that are authentic, on this account.

Other accounts of what it means to be an authentic individual or to lead an authentic life, however, appear not to be demanding enough. For example, Levy has stated:

> It is true that some of us choose to embrace, or to remain in, ways of life that are in some ways antithetical to the ideal of authenticity – we join monasteries, or adhere to religions that regulate every aspect of our lives, even to the point of deciding who we shall marry and what careers (if any) we shall have [...] even when we embrace ways of life that require us to cede control of our significant choices to others, we often justify our decision in ways that invoke authenticity: we find this way of life personally fulfilling; it is, after all, our way of being ourselves.
>
> (2007, pp. 74–75)

Which ways of acting would *not* reflect authentic decisions or ways of life, if all the lifestyles that Levy describes as 'in some ways antithetical to the ideal of authenticity' could in fact be regarded as authentic?

Even if we assume that choices that are imposed on us by external factors need not necessarily be inauthentic – for example, in cases where we identify with the choice we make, even where it is made as a result of external pressures – Levy's account does not seem to point to a way in which authentic life choices could be distinguished from inauthentic ones. How would we know whether a particular person's choice to cede control over a particular decision is authentic or not? If we want to maintain that authentic life choices are better than inauthentic life choices, the account of authenticity that we choose to adopt needs to be accompanied by a method for distinguishing acts that live up to this ideal from those that do not. The above accounts of what it means to be an authentic individual or to lead an authentic life proposed by Levy do not seem to offer a way in which to distinguish authentic decisions, individuals, or lifestyles from their inauthentic counterparts. It could be argued that this, in turn, should lead us to question the analytic usefulness of the notion of authenticity. It would seem that, given the absence of a means to distinguish the authentic from the inauthentic, the authenticity ideal has limited analytical usefulness in general, and is particularly unhelpful as a tool for distinguishing interventions that individuals should be permitted to undergo from those that they should be prevented from undergoing. Objections to ME on grounds that it could undermine the authenticity in our experiences or existence have been directed towards interventions that are classified as mood *enhancement*, rather than towards interventions that enhance mood only as part of accepted medical practice (e.g., the use of antidepressants by persons suffering from clinical depression). These objections therefore presuppose that a distinction between 'treatment' and 'enhancement' could reliably be drawn. However, this is in itself a contentious claim.

The distinction between treatment and enhancement is often taken to distinguish cases where intervention is justified or mandatory from cases where intervention is supererogatory, or may even, according to some commentators, be morally impermissible. Many commentators have questioned whether a distinction between treatment and enhancement can in fact be drawn and, even if it could, whether such a distinction would hold any moral significance (see, e.g., Bostrom and Savulescu 2009; Harris 1992).

In addition, interventions that target either cognitive capacities or mood may be an area where a distinction between treatment and enhancement could be the most difficult to draw. There are several reasons for this. First, in the case of interventions that target either cognitive capacities or mood, a single type of intervention could be considered medical treatment in some contexts and an enhancement technology in others. For example, it has been proposed that the use of SSRIs could improve the sense of well-being both in people who suffer from depression and those who do not; furthermore, stress inhibitors such as propranolol (beta blockers) could be used by healthy individuals as a means of dealing with stressful situations.

The pace at which the use of SSRIs and similar forms of interventions has increased over recent decades has been taken as a sign of a development frequently referred to as the 'medicalization of unhappiness' (see, e.g., Conrad 2007; Elliott 2007; Shaw and Wordward 2004), a proportion of which could constitute off-label use or 'mood enhancement'. As Robert Whitaker (2010) has observed, in the two decades following the 1987 United States Food and Drug Administration approval of what has become the most widely known antidepressant – Prozac – the number of Americans who qualified for welfare benefits as a result of their 'suffering from a mental disability' rose to 3.97 million. As Whitaker has noted, '[i]n 2007, the [...] rate was 1 in every 76 Americans [–] more than double the rate in 1987, and six times the rate in 1955' (2010, p. 7).

Those who object to ME tend to argue that this kind of medicalization of human unhappiness leads to inauthenticity in human existence or experience. For example, Elliott has advanced that,

> Some kinds of responses to the world are reasonable even when they are disturbing... For all the good that antidepressants do, there remains the nagging suspicion that many of the things they treat are in fact a perfectly sensible response to the strange times in which we live.
>
> (1999, p. 68)

Thus, the fear is that the blocking of 'perfectly sensible' responses to our environment through the use of ME may threaten our sense of what it means to be naturally human. We may forget what full flourishing or true human happiness really entails. Full human flourishing comprises more than mere happy feelings, and requires a connection to reality, as opposed to the unnatural or artificial improvement of mood that drugs bring about (Bolt 2007, p. 287).

The idea that full human flourishing requires something beyond the mere presence of feelings of happiness and, in particular, that it requires some kind of connection between mood and 'reality', resonates the idea on which Nozick's Experience Machine is based. The claim that the use of ME could threaten what it means to be naturally human, through severing the connection between the mood that we experience and 'reality', could be interpreted as a claim akin to 'externalism' in the philosophy of mind. On this view, the content of mental states depends on facts, some of which are external to the person experiencing them. Thus, although while plugged into the Experience Machine we could, for example, have the experience of successfully defending our PhD thesis, this experience would be different from the experience (outside the Experience Machine) of *in fact* successfully defending it. This would be the case even where these two experiences were perceptually or subjectively indistinguishable. The idea is that authentic experiences, that is, experiences of in fact doing something, are somehow more valuable than merely having the experience of doing something. However, if authentic and inauthentic experiences were subjectively indistinguishable, what could explain the inferiority of inauthentic experiences? Proponents of the authenticity ideal could argue that the quality of our experiences is not exclusively determined by their subjective content, but that their origin will also have an impact on their quality. This argument, however, seems simply to assume the conclusion for which it has set out to argue. The fact that Experience Machine-induced experiences are induced by the Experience Machine cannot in itself explain their inferiority – an additional reason is needed to justify the claim that they are inferior. Thus, even if we accept that the quality of an experience is determined both by its subjective content and its origin, we would need a further argument to convince us that the Experience Machine as an origin of experience is inferior to an 'authentic' origin.

Nozick's position could be understood as a claim that while the subjective quality of our experiences in the Experience Machine would be extremely high, intuitively we would still think something was lacking. However, Nozick fails to offer a plausible account of how we would know whether this something was missing. Once plugged into the Experience Machine, we would certainly be under the impression that nothing was missing from our lives – that is the very nature of the Machine. To the extent that we hold the intuition that something is lacking, this is because we *know* that life plugged into the Experience Machine would not be authentic. Applying the same line of reasoning to the question of the permissibility of ME, we would be committed to arguing that we would reject ME because we would know that our experiences under their influence would be inauthentic. This line of reasoning thus begs the question that it purports to answer.

Of course, it might be objected that the analogy between ME and the Experience Machine is somewhat 'unrealistic'. Using the former is unlikely to be as 'effective' as plugging into the latter. Unlike the Experience Machine, ME would not be likely to create hallucinations or illusions. It may enhance our mood, but it is unlikely to create completely new experiences that lack any connection to reality. Those who argue that the lack of connection to reality is what makes Experience Machine-induced experiences inauthentic would be likely to hold that ME was less problematic than the Experience Machine, since ME-induced experiences on this view would be less inauthentic. However, we would need a further reason that could justify the assumption that a connection to reality makes an experience superior to experiences that lack such a connection. Indeed, it could be argued that we do not even have a way in which to distinguish experiences that have a connection to reality from those that lack one. While plugged into the Experience Machine, we would not know that our experiences lacked a connection to reality. How would we know whether our ME-induced experiences had a connection to reality or not? How do we know if any of our experiences have a connection to reality? I would argue that if we cannot reliably distinguish between authentic and inauthentic experiences, it is nonsensical to assume that authentic experiences are superior to inauthentic ones.

## The usefulness of natural emotional responses to life events

The claim that there are reasons for preserving the connection between our mood and 'reality' may, however, instead be interpreted as a claim about the value of preserving natural emotional responses to life events. For example, it could be argued that the fact that serotonin levels rise and fall in response to life events suggests that interfering with them (in non-pathological circumstances) could be inadvisable. There is some empirical evidence to suggest that certain emotional responses to life events are, indeed, adaptive. For example, there is some evidence that mood helps people to adapt to the social positions in which they find themselves, or to the social status they are accorded in a particular situation, and that the capacity for such adaptations has thus conferred an advantage throughout the course of human evolution (See, e.g., Nesse 1991, p. 35; Nesse 1998, pp. 411–12). Since it could, for example, be maladaptive to have high self-esteem regardless of social opinion (since this could, among other things, cause such individuals to be excluded from the group) (Nesse 1991, p. 34), we may have reason to fear that widespread use of mood enhancers or anxiety relievers may interfere with the ways in which human social hierarchies are regulated. In addition, mood and emotional responses could serve important functions in the allocation of one's resources (Nesse 1998, p. 405, see also Bibring 1953; Brickman 1987; Carver and Scheier 1990; Klinger 1975). In fact, it has been suggested that people in general are disposed to overestimate and be overly optimistic about their own capacity and skill, the degree of control they have over their life and environment, and about how well things are likely to go for them in the future (Taylor 1989; Weinstein 1980; Weinstein 1984). Such responses are valuable in that they encourage us to persist in enterprises that may be temporarily unprofitable but which are likely to pay off in the longer term (Nesse 1991, p. 35). Negative feelings also serve important functions in the allocation of the resources that individuals have at their disposal. As Randolph Nesse has observed, '[l]ife's important decisions are usually questions about whether to maintain the status quo or whether to change the pattern of resource allocation' (1991, p. 34). Negative feelings or low mood encourage us to withdraw resource investments from enterprises that are likely to be futile. These responses occur, for example, when our efforts at some particular enterprise repeatedly fail. As psychiatrist Emmy Gut has observed, depression could often arise in individuals who are faced with a failing primary life strategy, and where no alternative life strategies seem readily available. In such cases, depression

> may permit reevaluation and insights that can solve the puzzle of the inner deadlock. Our initial striving toward a goal can then be understood and implemented with better methods; the goal can be revised, or else abandoned if recognized as unrealistic.
>
> (1985, p. 99)

Some types of suffering or low mood may thus form part of evolutionary mechanisms developed to help us survive in our environment. Does the fact that 'natural' responses to life events may be adaptive imply that we ought not to interfere with them, for example through the use of ME? As Nesse has observed, not necessarily:

> If the mechanisms that regulate the emotions are products of evolution, it might seem to follow that interfering with them will usually be unwise. After all, natural selection has had millions of years to shape the mechanisms, and so by now their thresholds should be set to near-optimum levels. But everyday medical practice contradicts that conclusion. People routinely take aspirin for pain and fever with few untoward consequences; anti-nausea medications relieve much suffering with only occasional complications; ten million Americans each year take anxiety medications, yet there is no epidemic of risky behavior.
>
> (1991, p. 37)

The fact that the routine interference into people's natural responses to life events that modern healthcare provision represents has not led to a backlash of negative consequences may suggest that many of our 'natural' responses to life events could be over-responsive (Nesse 1991, p. 37). One of the main reasons for this over-responsiveness could be that these responses may reflect not the society in which we live today, but the one for which they were first developed. From an evolutionary perspective, there may be several reasons why our 'natural' responses are over-responsive in the context of contemporary society. First, the ways in which humans have been 'programmed' to respond emotionally to life events do not change as rapidly as our societies have changed in recent centuries. Second, there are inherent mechanisms in evolutionary 'programming' that cause the threshold for these kinds of responses to be set low, that is, 'easily activated'. The reason for this is that the consequences of a failure to evoke the right kind of defense-response (e.g., the failure to evoke a flight response when coming across a hungry tiger) would, in the course of evolution, have been much more detrimental than the costs of over-responsiveness (Nesse and Williams 1995). As Nesse has noted, '[i]f much suffering is unnecessary, there should be many occasions on which it can be safely blocked' (1991, p. 37). Thus, from the premise that many of our emotional responses may be useful reactions to life events, it does not follow that interference with such responses, for example through the use of ME, is always inadvisable. Instead, it may suggest that blocking emotional responses may *sometimes* be counterproductive. It could be advisable to avoid using ME to block useful forms of emotional responses, for example, with regard to guiding our resource investments in different life projects. However, we would not have reason to preserve *all* 'natural' emotional responses on the grounds that they are more natural or more 'authentic' than responses that are induced through the use of ME. Perhaps more research is needed into the identification of situations where emotional responses could safely be blocked *vis-à-vis* situations where negative feelings are useful and their blocking may thus pose risks. It may seem reasonable to assume that it would be counterproductive to block at least some kinds of emotional responses. This would not, however, be on the basis of concerns over authenticity, but on the basis of its likely consequences for our long-term well-being. Moreover, if we justified non-interference by reference to the fact that emotional responses to life events are *natural,* how could we justify interference to stop natural (or evolutionary) processes in other cases, through the provision of medical treatment? As John Harris has observed, 'the whole practice of medicine might be described as a comprehensive attempt to frustrate the course of nature' (1985, p. 38). We, as a society, may have *other* reasons to prohibit the use of (certain kinds of) ME, for example, if widespread use would undermine the functioning of society itself. For example, if certain forms of ME would prevent people from doing things that are valuable to society, such as having jobs, performing acts of solidarity, through, as per the Experience Machine, instilling in us the *experiences* of having jobs and performing acts of solidarity, when in fact we were floating around in a tank with electrodes attached to our brain, there might be good reasons for prohibiting the use of these particular forms of ME.

Indeed, it could be argued that what distinguishes plugging into the Experience Machine from using ME is that in the latter case, the people around a mood enhanced individual, not all of whom are themselves using ME, could be negatively affected, Nozick stipulates that this is not the case in the Experience Machine thought experiment.

Another way in which ME could potentially negatively affect the functioning of society is through contributing to a process of 'individualization of suffering' (Levy 2007, p. 129). A constant search for quick solutions, and pharmacological solutions in particular, may, as Levy has pointed out, be politically regressive. It may steer attention away from societal or political problems, and discourage us from seeking social or political solutions (Ibid.). For example, we could imagine a scenario where concentration-enhancing drugs (such as methylphenidate) were used to deal with stressful classroom situations, or where individuals working in certain professions used wakefulness promoters (such as modafinil) to be able to cope with unreasonable working hours. A turn to pharmacological symptom relief in such cases could prevent the underlying political or social issues from being addressed.

These kinds of potential societal effects of widespread use of ME may be a compelling reason to restrict their use. Reasons based on negative effect for the functioning of society of widespread use of ME would, however, be consequentialist in their nature, and could be explained without any reference the authenticity ideal. Neither does the application of the authenticity ideal seem to pick out any of the reasons that could justify restrictions of the use of ME. To this end, the authenticity concept does not seem to do the work that opponents of ME would like it to do.

## Conclusion

If we found ourselves in a situation where we were forced to choose between greater happiness for ourselves on the one hand, and a life that was 'authentic' on the other, would we have a good general reason to choose the latter over the former? Nozick argued that we should not plug into the Experience Machine, even though it would be certain to give us all the experiences we desired, because a life lived in it would lack the necessary connection to reality that is required for a life to be 'authentic'. It has been suggested that this reasoning could also be applied to ME. ME should be avoided because its use risks rendering our experiences – or indeed our very existence – less authentic.

In this paper, I have argued that neither positions that invoke the Experience Machine, nor those according to which natural responses to life events should be preserved, demonstrate convincingly that ME is undesirable or impermissible. In some cases its use could be inadvisable, but in others it may be both permissible desirable.

Both arguments against ME on grounds of its negative effects for authenticity examined here, fail to present a way in which we can distinguish cases where ME is impermissible from cases where it could be permissible (if there are such cases). Instead, they appear merely to assume the conclusion that they set out to prove; that is, that authenticity is something we ought to value and strive for in our lives. Indeed, they seem to argue not only that we should value authenticity, but that we should (sometimes) value it over happiness, if faced with a choice similar to that of whether or not to plug into the Experience Machine. This choice may not be as moot as it first seems, for, as South African philosopher David Benatar has noted,

> [t]here is quite a bit of evidence that happier people with greater self-esteem tend to have a less realistic view of themselves. Those with a more realistic view tend either to be depressed or have low self-esteem, or both.
>
> (2006, p. 65)

If it is indeed the case that we might – to some extent – have to choose between a greater degree of happiness and a greater degree of authenticity (or realism, in Benatar's words), I suggest that the onus would be on those who argue that we ought to value authenticity over happiness to explain why this is the case. In this paper, I have considered two arguments against the use of ME. Both of these arguments claim that ME somehow undermines the authenticity of human experience or existence, and that this is somehow objectionable. Both arguments rely on the assumption that a connection to reality is necessary for an experience to be authentic, and that the lack of such a connection to reality is problematic. According to the first argument, authentic experiences are *qualitatively superior* to inauthentic ones. I have suggested that this argument fails because it does not provide a way in which to distinguish authentic or authenticity-promoting experiences or choices from inauthentic or authenticity-undermining experiences or choices, and because it does not show *why* authentic experiences would be qualitatively superior to inauthentic ones. According to the second argument, authentic or natural emotional responses to life events are useful, and ought therefore not to be interfered with. On this view, ME is wrong because it interferes with natural (authentic) emotional responses to life events. I have proposed that this argument fails because it does not succeed in showing that it is always wrong to interfere with natural responses to life events. In addition, it does not pick out what could be compelling reasons not to interfere with natural emotional responses to life events.

Here, I have argued that both arguments fail to capture (what may be) compelling reasons to restrict the use of (some forms of) ME. In neither case, the authenticity concept seems to do the work that opponents of ME would like it to do. The only compelling reasons to restrict the use of ME are consequentialist reasons relating to potential societal effects of widespread use. Thus, there may indeed be reasons for us, as a society, to restrict the use of (some forms of) ME. Its potential effect on the authenticity of human experience, however, is not one of them.

## References

Benatar, D. (2006): *Better Never to Have Been. The Harm of Coming Into Existence.* Oxford: Clarendon Press.

Bibring, E. (1953): The mechanisms of depression. In: Greenacre, P. (ed.) *Affective Disorders.* New York: International Universities Press, pp. 13–48.

Bolt, L. L. E. (2007): True to oneself? Broad and narrow ideas on authenticity in the enhancement debate. In: *Theoretical Medicine and Bioethics*, 28(4), pp. 285–300.

Bostrom, N. and Savulescu, J. (eds.) (2009): *Human Enhancement.* Oxford: Oxford University Press.

Brickman, P. (1987): *Commitment, Conflict and Caring.* Englewood Cliffs, NJ: Prentice-Hall.

Carver, C. S. and Scheier, M. F. (1990): Origins and functions of positive and negative affect: A control-process view. In: *Psychological Review*, 97(1), pp. 19–35.

Conrad, P. (2007): *The Medicalization of Society: On the Transformation of Human Conditions into Treatable Disorders.* Baltimore: Johns Hopkins University Press.

Elliott, C. (1999): *A Philosophical Disease: Bioethics, Culture and Identity.* New York: Routledge.

Elliott, C. (2007): Against happiness. In: *Medicine, Health Care and Philosophy*, 10(2), pp. 167–171.
Elliott, R., Sahakian, B. J., Matthews, K., Bannerjea, A., Rimmer, J. and Robbins, T. W. (1997): Effects of methylphenidate on spatial working memory and planning in healthy young adults. In: *Psychopharmacology*, 131(2), pp. 196–206.
Harris, J. (1985): *The Value of Life: An Introduction to Medical Ethics.* London: Routledge.
Harris, J. (1992): *Wonderwoman and Superman: Ethics of Human Biotechnology.* Oxford: Oxford University Press.
Klinger, E. (1975): Consequences of commitment to and disengagement from incentives. In: *Psychological Review*, 8(2), pp. 1–25.
Kramer, P. (1993): *Listening to Prozac.* New York: Viking.
Levy, N. (2007): *Neuroethics. Challenges for the 21st Century.* New York: Cambridge University Press.
Nesse, R. M. (1991): What good is feeling bad? The evolutionary benefits of psychic pain. In: *The Sciences*, 30, pp. 30–37.
Nesse, R. M. (1998): Emotional disorders in evolutionary perspective. In: *British Journal of Medical Psychology*, 71(4), pp. 397–415.
Nesse, R. M. and Williams, G. C. (1995): *Why We Get Sick: The New Science of Darwinian Medicine.* New York: Times Books.
Nozick, R. (1974): *Anarchy, State, and Utopia.* New York: Basic Books.
Shaw, I. and Wordward, L. (2004): The medicalisation of unhappiness? The management of mental distress in primary care. In: Kauppinen, K. and Shaw, I. (eds.) *Constructions of Health and Illness: European Perspectives.* Aldershot: Ashgate.
Taylor, S. H. (1989): *Positive Illusions: Creative Self-Deception and the Healthy Mind.* New York: Basic Books.
Weinstein, N. D. (1980): Unrealistic optimism about future life events. In: *Journal of Personality and Social Psychology*, 39(5), pp. 806–820.
Weinstein, N. D. (1984): Why it won't happen to me: Perceptions of risk factors and susceptibility. In: *Health Psychology*, 3(5), pp. 431–457.
Whitaker, R. (2010): *Anatomy of an Epidemic.* New York: Broadway Paperbacks.

# 9

# Prometheus Descends: Disabled or Enhanced? John Harris, Human Enhancement, and the Creation of a New Norm

*Trijsje Franssen*

## Introduction: Questioning human enhancement

One of the most important questions in the ethical discussion on human enhancement is whether enhancement (by a certain definition) is acceptable in the first place, and if so, whether we have a moral duty to enhance people when possible. Some scholars employ a concept of 'normalcy' in order to draw a distinction between therapy and enhancement, establish what counts as a disease or disability, and determine which treatments are morally acceptable or obligatory ('therapeutic' ones, which bring a person back to his or her normal state) and those which are not ('enhancing' ones, which improve upon a normal, healthy life). The pro-enhancement bioethicist John Harris, however, claims that the concept of 'normalcy' is morally irrelevant, and that there is no moral difference between enhancement and therapy. In an early paper, he defines disability as a 'physical or mental condition we have a strong rational preference not to be in [...], a condition which is in some sense a "harmed condition"' (1993, p. 180). Though Harris acknowledges that his definition is 'necessarily vague', he argues that this is not a problem since 'we know what injury is and we know what disability or incapacity is' (Ibid.) – a condition that might shorten a person's lifespan or make him or her particularly vulnerable to infection, for instance.

Harris explains that a disease or disability is relative to 'possible functioning': the possible alternative conditions at a given point. What some might characterize as a 'normal state' could therefore also be a 'harmed condition', and so we have a moral duty not only to cure disability, but also 'a positive moral duty to enhance' (2007, p. 3):

> The overwhelming moral imperative for both therapy and enhancement is to prevent harm and confer benefit. Debated in that moral light, it is unimportant whether the protection or benefit conferred is classified as enhancement or improvement, protection or therapy. (Ibid., p. 154)

In this paper I criticize Harris' definition of disability from different angles. First, I argue that the definition does not seem to meet his own requirements. He claims that his definition of disability is independent of both the feelings of the subject of the definition and the surrounding social conditions. Harris nevertheless characterizes it by phrases such as 'the deprivation of worthwhile experience', or the fact that 'pleasures' are closed to a person – characterizations that are hard to determine without reference to a person's feelings. Furthermore, if disability is relative to 'possible functioning', it seems impossible to maintain that the concept is indeed independent of social and cultural conditions. Second, Harris' emphasis on our preference not to be harmed as being rational can be questioned as well. The concept of rationality is not as straightforward as he may want it to be, and it automatically classifies disabled people who are at peace with their condition and who would not prefer to be without it as irrational. Third, I argue that accepting his definition would have some radical and strange consequences. Harris claims (among other things) that enhancement has always been a part of human evolution. However, he hereby seems to reintroduce a concept of normalcy: what used to be an ideal – the enhanced human being – appears as the new norm, even though in his definition he extensively argues that normalcy is morally irrelevant.

In order to show the pervasiveness of Harris' (and other similar) ideas in the enhancement debate, I will discuss the writings of some advocates of enhancement who all compare the mythological figure of Prometheus with the enhanced human. These passages show that what was once an ideal has, for them, become the standard for the true human. However, if the enhanced human becomes the norm – which Harris and others seem to maintain – then the non-enhanced human becomes a dysfunctional being. One of the consequences of this would be that the vast majority of human beings alive today would become subhuman or even disabled, which raises important questions about disability and about the practical implications of this situation with regard to social relations, legal rights, and equality.

## John Harris in the human enhancement debate

The current debate on 'human enhancement' centers on the question of whether or not we should try to 'enhance' or 'improve' the human being by means of – mostly emerging – technologies, such as reproductive technologies, cloning, and genetic engineering. Based on a variety of arguments, theorists in the pro-enhancement camp argue that such improvement should be pursued. These technologies will, they claim, cure disease, reduce inequality, make us physically and mentally stronger, bring us happiness, and, according to some, in the end even generate a whole new species. This 'posthuman' species will, as famous proponent (and leading transhumanist) Nick Bostrom puts it, have physical and mental capacities 'greatly exceeding the maximum attainable by any current human being' (2008, p. 107). The other camp, baptized 'bioconservatives' by their opponents, are skeptical of the effects that supporters claim human enhancement will have. Ranging from practical doubts with regard to its feasibility to passionate condemnation of any possible application of such technologies to humans, these skeptics argue against enhancement. It will not make us happier, stronger, and more equal, they say. On the contrary, it threatens our dignity and autonomy, and could promote injustice or even dehumanize us.

Part of the discussion focuses on the distinction between therapy or treatment on the one hand and enhancement on the other. This distinction is made or referred to in order to answer the question of which interventions into or alterations of human functioning are morally acceptable or (socially) obligatory and which are not.

Skeptics of enhancement often employ some kind of concept of 'normalcy' or 'species-typical'[1] functioning in order to draw a line between (good) therapy and (questionable) enhancement. Bioethicist Norman Daniels, for instance, argues that the obligations of a just healthcare service consist of the prevention and treatment of disease and disability, which are defined in the following way:

> Disease and disability, both physical and mental, are construed as adverse departures from or impairments of species-typical normal functional organization, or 'normal functioning' for short.
>
> (Daniels 2001, p. 3)

Therapy, in other words, means to cure disease, it restores 'normal functioning'. Enhancement, on the other hand, entails improvement upon a normal, healthy life: 'In effect, anything to do with maintaining normal function falls under the scope of "treatment" as opposed to enhancement' (Daniels 2009, p. 34). This means that while we do have a moral obligation to cure disease, we do not have such an obligation with regard to enhancement. And although this does not mean that enhancement should be rejected in principle, Daniels does think that we should be very careful, for 'if we are trying to improve on an otherwise normal trait, the risks of a bad outcome, even if small, outweigh the acceptable outcome of normality [...]. I believe this argument has great force' (Ibid., p. 38).

Others occupy an even stronger position and passionately dismiss whatever goes beyond the 'normal' state. Speaking of genetic engineering in particular, W. French Anderson argues that

> a line can be drawn and should be drawn to use gene transfer only for the treatment of serious disease and not for any other purpose. Gene transfer should never be undertaken in an attempt to enhance or 'improve' human beings.
>
> (1990, p. 21)

This is so because, as Anderson formulates elsewhere, genetic intervention that brings one beyond the normal state is 'fraught with danger' (1994, p. 759).[2]

Unsurprisingly, those in favor of human enhancement take a very different position with regard to healthcare's moral obligations. John Harris, for instance, argues that instead of prohibiting enhancement or drawing a line between therapy and enhancement by means of a concept of normalcy, we have a moral duty to enhance. Harris claims that one cannot meaningfully distinguish between enhancement and therapy in moral terms, and criticizes the concept of normalcy as employed by Daniels and others. He gives several reasons to support this claim.

- First, what makes up a 'normal healthy life' changes all the time, and 'is determined in part by technological and medical and other advances' (Harris 2007, p. 93). It is common today (in developed countries) to be protected against smallpox, for instance, though this was not always so. Furthermore, in a couple of decades we might consider it normal to have a life expectancy of 120 years, while we obviously do not today.

- Second, what is generally considered to be 'abnormal' may still be morally acceptable or even obligatory. Harris pictures a new genetic intervention that would make us invulnerable to HIV, and claims that he finds it hard to imagine that it would not be made 'imperative' instead of merely 'permissive' because of its allegedly 'abnormal' character; that despite its 'abnormal' gene-manipulating qualities such an enhancement would not be embraced. Similarly, we see that today, when it comes to vaccination, '[i]nterestingly, there has been very little resistance to this form of enhancement' (Ibid., p. 21). Indeed, according to definitions of normalcy (such as Daniels'), vaccination does count as an enhancement technology, since immunity to measles, for example, is not part of our 'normal functioning'. The case of vaccination demonstrates, Harris argues, that a health service's moral obligation is thus not dependent on an intervention's 'normalcy', or on it merely being therapy.
- Third, what is considered to be perfectly 'normal' might also be unacceptable – painful birth giving, for instance, or the diseases of old age. If we developed treatments for the latter and this led to substantial life extension or even immortality, it would, Harris argues, seemingly entail both therapy and enhancement. Yet these treatments would only appear to be therapeutic – and thus 'acceptable' – 'because treating disease seems typical of therapy, not because normal species functioning does or can play any role at all in the argument' (Ibid., p. 45). According to Harris it shows that – as does Daniels' argument – a trait or treatment's acceptability does not depend on the concept of normalcy. In fact, if one would strictly employ Daniels' definition of disease as being relative to normalcy, treating the diseases of old age would be abnormal, enhancing, and thus unacceptable. This, however, is very unlikely to be a statement Daniels would agree with.
- Fourth, the exact same therapy that restores one person's 'normal function' could actually be enhancing for others. Harris envisions a form of treatment for people with brain damage, which could be radically enhancing for those with a 'normal' cerebellum. Treating brain damage on the basis of stem cell regeneration, for instance, could in turn be conceived of as enhancement if applied to a 'healthy' brain's capacities; similarly, 'therapy' for amnesia could enhance a 'normal' brain's memory.

In short, according to Harris the concept of normalcy is morally insignificant. Whether a disease or treatment is labeled as 'normal' is context and history dependent. The same goes for enhancement and therapy: what is considered to be an 'enhancement' today might be 'therapy' tomorrow, and what counts as mere treatment for one person could be radically enhancing for another. The distinction between enhancement and therapy cannot be made by reference to normalcy; normalcy has no ethical (or explanatory) significance. Therefore, our moral reasons to treat people should not depend on any of these three concepts. It is interesting to note here that the same argument is also applicable to appeals to nature, or to the 'natural', when it comes to determining what is morally right. It is natural for people to fall ill, for instance, and so '[w]hat is natural is morally inert and progress dependent' (Ibid., p. 35).

As an alternative, Harris claims that an intervention's moral acceptability or our obligation to perform it should depend on the 'harms this will prevent and the goods that this will bring about' (Ibid., p. 54). Since enhancements are, by definition, 'obviously good for us' – they are 'an improvement', 'a difference for the better', and they protect 'the safety of the people' (Ibid., pp. 35, 131, 36, 151), and so on – people should of course be entitled to use them, and in the end, perhaps even be obliged to do so. To protect from harm is what matters, and both the failure to cure someone and the failure to protect someone by means of enhancement (for instance, from a future disease or an 'unnecessary death') means to harm them. Therefore, there is 'no moral difference between attempts to cure dysfunction and [...] to enhance function' (Harris 1993, p. 184): the therapy–enhancement distinction should be abandoned.

In accordance with his emphasis on the prevention of harm, Harris defines disability as a 'physical or mental condition we have a strong rational preference not to be in [...], a condition which is in some sense a "harmed condition"' (Ibid., p. 180). To clarify what he means, he asks how someone who feels that she has become disabled as a consequence of 'industrial effluent' would convince a court that she has been injured. 'The answer is obvious but necessarily vague. Whatever it would be plausible to say in answer to such a question is what I mean [...] by disability and injury' (Ibid.). If the woman has become particularly vulnerable to infection due to the factory's waste, says Harris, or if her lifespan has been limited by it, she would surely be acknowledged as having been injured, and perhaps even disabled. Similarly, if as standard almost everyone has their life expectancy prolonged because they have been protected against infections from birth, 'it would surely be plausible to claim that failure to protect [a child at birth] [...] constituted an injury and left them disabled' (Ibid.).

Obviously, according to Harris, disease or disability cannot be 'defined relative [...] to normal species functioning' (2007, p. 53), since, as said, the 'normal' changes according to time and context. Rather, they should be defined 'relative to possible functioning' (2009, p. 150): what counts as 'disabled' or 'dysfunctional' depends on the possible alternative conditions at a given point. If available (if 'possible'), if the gains are important enough and the risks are worth taking, we are thus morally obliged to use enhancements that improve our functioning – even if this means changing human nature, for 'changes in human nature [...] seem ethically uninteresting' (Ibid., p. 136). If we fail to enhance human functioning where possible, this is the same as failing to cure disease or disability. If there was a gene therapy, for instance, that could cure some genetic disorder, 'to fail to use it would be *deliberately to harm* those individuals' (1993, p. 183; my emphasis). Similarly, to fail to enhance a person's life expectancy by means of genetic engineering would be to harm him.

## Criticism: Socio-cultural factors and Harris' reintroduction of normalcy

There are several aspects of Harris' definition of disability which I would like to discuss. First, as we have seen, Harris does not define disability with respect to normalcy because of the ambiguity and relativity inherent to the concept. Moreover, he claims that his definition avoids the 'pitfall' of depending on '*post hoc* ratification by the subject of the condition – it is not a prediction about how the subject of the condition will feel' (Ibid., p. 181); it can therefore also be used for the 'temporarily unconscious', embryos, and so on. Harris seemingly wants to avoid a definition that is dependent on subjective feelings or suffering, and instead aims to define disability in an objective way: depending on facts and rational judgment (I will return to the second aspect below).

In response to Newell (1999) and Reindal (2000), two scholars who criticize Harris (or ideas similar to his) for ignoring the social and cultural factors that turn an impairment or condition into a problem – and hence a disability – in the first place, Harris claims that these factors really do not play a role when it comes to an impairment being a disability. 'The harm of deafness', Harris argues, 'is the deprivation of worthwhile experience' (2000, p. 98). Social factors can definitely aggravate the problem of having a disability, but disabilities:

> are disabilities because there are important options and experiences that are foreclosed by lameness, blindness and deafness. There are things to be seen, heard and done, which cannot be seen, or heard, or done by the blind, deaf and the lame *whatever the social conditions*. [...] [T]here are pleasures, sources of satisfaction, options and experiences that are closed to them. In this lies their disability.
>
> (Ibid.)

Although Harris acknowledges that disabled people do experience social problems, he claims that 'these are separate sorts of harms although, of course, they are causally related' (Ibid.). These are problems we should be capable of dealing with 'independently of whether or not they are triggered by disability. Hence they are not a definition or conception of disability but part [...] of what is *bad* about disability' (Ibid.).

In other words, the concept of disability is not determined as such by social and cultural conditions, nor does it, as Harris emphasizes, merely depend on a person's – or, in his wording, 'the subject of the condition's' – feelings. Rather, it depends on the 'deprivation of worthwhile experience', on the fact that 'pleasures' and 'sources of satisfaction' are closed off to him or her. However, how can these characterizations be independent from either a person's feelings or the social context? Even if Harris, as it seems, wants to focus on the objects and the sources that cause these pleasures and satisfactions, it is very unlikely that they are completely independent of context. Of course one could argue that however you interpret the world, sounds simply exist and deaf people do not hear them. It might therefore seem reasonable to claim, as Harris does, that deafness, blindness, lameness, and so on do indeed involve some loss of capacity.

However, as philosopher of biology John Dupré states, when analyzing (dis)abilities, 'it is essential to distinguish between intrinsic capacities and relational capacities' (1998, p. 229); that is, between the purely neurological or physiological abilities, and the ones which depend just as much on the environment, technological devices, and other instruments. For in general it is not merely the intrinsic ground of a capacity but also the elaboration of a capacity in context that matters. Thus if those who cannot hear see no problem in their deafness –

which quite a number of deaf people claim – and/or if every public establishment were practically adapted to deaf people – interpreters everywhere, everyone articulating clearly and thus allowing for easy lipreading– it would be quite hard to continue to characterize deafness as a disability. Actually, many deaf people view themselves as members of 'the Deaf community' – as just another minority rather than a disabled group – and argue one should use a capital D when referring to 'the Deaf'. In short, whether something can and should be characterized as a disability can be seen as largely dependent on environmental or socio-cultural factors. It is therefore too limited to focus, as Harris does, exclusively on the intrinsic ground of a capacity when classifying a disability.

Beyond the limited focus of this argument, Harris also seems to contradict himself, for if disability is relative to 'possible functioning' and available alternative conditions, it seems impossible to maintain that the concept is indeed independent of socio-cultural conditions. As he emphasizes when criticizing the concept of normalcy, the possible alternatives (whether practical or theoretical) for somebody with a particular condition are dependent on technological and medical developments, on a country's health system, its political situation, and so on. Harris thus defines disability, on the one hand, with respect to possible functioning, which is determined in a large part by socio-cultural and environmental factors, while claiming that it is independent of social conditions on the other.

A similar difficulty arises in his concept of 'harm'. Harris seems to hold that 'harm' or a 'harmed condition' is something objective that exists apart from social circumstances: there are simply 'things to be seen, heard and done, which cannot be seen, or heard, or done by the blind, deaf and the lame' (these are what Dupré refers to as intrinsic capacities). But at the same time, he characterizes harm – and hence disability – as 'the deprivation of a worthwhile experience', and emphasizes how enhancement is something that is 'good for us', 'an improvement', confers 'benefits', 'protects the safety of the people', and so on. However, the significance of words such as 'worthwhile', 'satisfactory', or simply 'better' or 'welfare', cannot be established independent of social, (sub)cultural, and historical circumstances or discourses. What was understood by 'welfare' in the Victorian era is quite different from what we consider to be welfare in the 21st century.

Taking Harris' own arguments with respect to normalcy into consideration, it seems as if he would agree with me on this point. As said, he claims that what counts as 'normal' or 'natural' changes all the time.

Furthermore, he argues that we would characterize treatments for the diseases of old age as therapy 'only because treating disease seems typical of therapy' (2007, p. 45). What counts as therapy, then, apart from being a contextual issue, is also a semantic question. The fact that we characterize something as therapy largely depends on the meaning of the word and its daily use, namely as something that includes the treatment of diseases. Hence, even when taking (some of) Harris' own claims into account, it seems more likely that words such as 'worthwhile' and 'welfare' are context, culture, and discourse dependent. Since Harris defines harm with respect to these words, he contradicts himself when he claims simultaneously that disabilities are disabilities 'whatever the social conditions'.

The second aspect of Harris' definition that I want to discuss is his emphasis on disabilities as conditions that we have 'a strong *rational* preference not to be in' or 'which a *rational* person would wish to be without' (1993, p. 180; 2000, p. 98; my emphasis). This is a significant detail, since each time Harris evokes his definition he makes sure that the notion of rationality is included, in order to underline the fact that harm and disability are not determined by social conditions but instead by biological and technological conditions. This emphasizes his definition's alleged neutrality. Purportedly, a rational judgment is a more objective judgment than an irrational one (or one based on emotions), the latter of which could lead someone to want to be disabled or harmed. However, what counts as rationality or a rational person in the first place is – just as with harm – dependent on social circumstances and cultural premises, which makes the definition less objective than Harris claims it to be. Furthermore, even if there was the possibility of an unambiguous understanding of what rationality is and who is classified as a rational person, several questions remain.

Imagine the scenario of a person – whom we (as rational persons) consider to be irrational but not disabled – who chooses to classify herself as disabled. Should this person's self-perception – or any of her perceptions – be completely discounted for the very reason that we consider her to be irrational (but nevertheless able-bodied)? Or what if a person – whom we would describe as disabled – does not consider himself as such? Would that make him – or at least his self-perception – irrational? Let us return to the issue of deafness. In 2001, a deaf, lesbian couple in the United States deliberately used sperm from a deaf friend in order to raise their chances of conceiving a deaf baby. Sharon Duchesneau and Candy McCullough succeeded; the baby was born congenitally deaf. One of the important reasons they gave for their choice was that they do not see deafness as a disability, but rather as an identity. They argued that they would even be able to be better parents to the boy, since they could talk to him in their own (sign) language, and introduce him to the 'Deaf community', of which they were contented and self-confident members themselves.[3]

The couple's action could be judged by some as being very selfish. However, the case provides an excellent example of the fact that membership of the 'Deaf community' provides many deaf people great satisfaction. It is experienced as a culture, with its own language, history, and way of life.[4] Clearly, in this case, social factors determine in large part whether a deaf person's condition is experienced as disabled or not. The fact that the person's possible functioning or enhancement possibilities seem worthless to her or him demonstrates that these too are value-laden and cannot be considered to be indisputably beneficial, as Harris suggests. The unharmed condition is therefore not necessarily the most rational – or the only rational – option. Choosing for a harmed condition – as the two deaf mothers in the case study above did for their child – may well be rational, if we accept (for now) the *Oxford English Dictionary*'s definition of 'rational' as meaning 'based on or in accordance with reason or logic' or being 'able to think sensibly or logically' (2003, p. 1461). If participation in such a community brings a person such (social) goods, it is very logical that these should be taken into account when determining her or his 'rational' preferences. This shows that what counts as rational is far from objective, that it can be reasonable to prefer a harmed condition, and, once again, that what counts as 'disabled' is not, as Harris claims, independent of social conditions.

The assertion that a person is irrational if he or she wishes to be harmed is simple to counter, since it is easy to provide examples of people who choose to be harmed for what would generally be considered legitimate and valid reasons; a soldier putting himself in the line of fire in order to save a fellow soldier, for example, or the courageous and altruistic workers who went to the site of the Fukushima Daiichi nuclear power plant in Japan after the damage caused by the March 2011 tsunami, exposing themselves to dangerous levels of radiation. Should these people be judged as irrational, or of doing irrational things, because they choose to be in a situation that exposes them to harm 'which a rational person would wish to be without' (Harris 2000, p. 98)?

All of these critical points touch upon Harris' treatment of the concept of normalcy. First, by claiming that the sole fact that the deaf, blind, and lame lack some options and experiences establishes their disability, Harris actually reintroduces a certain concept of normalcy: there is some range of capacities without which you simply are different, for lacking these makes you harmed – disabled – regardless of the (social) circumstances. Second, by replacing Daniels' (species-typical functioning) definition of disability with that of a 'harmed condition' that 'a rational individual would strongly desire not to be in', Harris suggests that the concepts of harm and rationality avoid the problems of the concept of normalcy, presupposing that they possess some objectivity that normalcy does not have. This assumption establishes both the unharmed condition and rationality as standard: it is seen simply as a fact that people do not want to be harmed, and that a normal person makes rational decisions. In other words, these 'facts' are actually value-laden, or statements that include notions of normativity: being unharmed means being in *good* health and a rational decision is a *good* one. To summarize, with his definition, Harris implicitly presents the unharmed, rational, and full-range-of-capacities-possessing person as the norm, thereby reintroducing a concept of normalcy and rendering his argument vulnerable to his own criticism.

The third problem I would like to discuss is related to what would happen if we were to accept Harris' definition of disability. I propose that it would have some quite extreme consequences, since if radical enhancement were possible:

- those who were not enhanced would, according to Harris, be disabled, because they would be in a 'harmed condition' with respect to the possible alternative states;
- those who would not enhance others – parents not enhancing their children, for instance – would be guilty of harming them;
- those who chose not to enhance themselves would be harming themselves.

Since this is quite a complicated question, I will address the second half of my paper largely to this issue.

The enhanced human – those 'healthier, fitter, and more intelligent individuals' or 'better people' (Harris 2007, p. 2) – at this time might seem to be merely a futuristic image, an ideal. But is it? According to Harris we already use enhancement technologies, and even something such as life extension by means of genetic engineering is no longer merely a utopian fantasy.[5] For Harris, the enhanced human is anything but an unattainable standard of perfection:

> I point out the continuity that exists between therapy and enhancement, the fact the [sic] human enhancement has always

> been both a conscious and unconscious part of human development and of evolution, and I underline the familiarity of the multifarious attempts we humans have made not only to better ourselves in the sense of improving our material circumstances and well-being, but literally to better ourselves.
>
> (Ibid., p. 4)

By claiming that humans have always tried to enhance and 'better' themselves somehow, Harris seems to establish human enhancement as normal – vaccination, for example, is 'of course an enhancement technology and one that has been long accepted' (Ibid., p. 21). In spite of his contempt for the concept of normalcy, however, Harris' arguments actually result in a new concept of normalcy: going beyond merely presenting the unharmed, rational, and fully capable person as normal, his logic implies that the ideal – the enhanced human – is the norm. Apparently, instead of being an unattainable ideal, the enhanced – or enhancing – human has become the norm, the standard we must live up to.

Similar ideas are also found among other pro-enhancement theorists, whereby what one might think of as an ideal picture of the human is suggested to be the norm. In the following I would like to illustrate this by discussing some of the ideas of three pro-enhancement theorists who have found a new standard in a classic example of an ideal: the mythological figure of Prometheus.

## The Promethean standard

One of these enhancement advocates is the biophysicist Gregory Stock. Stock argues that biological enhancement can no longer be stopped, nor should it. Some people, he says, think that because of the dangers that enhancement technologies may bring – from cheating within sports to designing our children or even destroying the human species – we should refrain from manipulating human genetics. He himself, however, is sure that it will happen anyway, for 'when we imagine Prometheus stealing fire from the gods, we are not incredulous or shocked by his act. *It is too characteristically human*' (2003, p. 2; my emphasis). Stock seems to believe that we will not be able to stop human enhancement because we are all like Prometheus.

Prometheus was a clever Titan, a god belonging to the race of deities that ruled before the Olympians. In his version of the myth, tragedian Aeschylus relates how Zeus conceived a plan to wipe out the whole human race. Prometheus took pity on the mortals, stole fire from heaven and gave it to them. He also granted them wisdom and taught them various techniques and arts. When Zeus found out, Prometheus was severely punished. Zeus chained the immortal, rebellious god to the peak of Mount Caucasus. There, a vulture would daily eat his liver, and every day his liver was renewed, resulting in eternal punishment. Many centuries of this torment passed before Heracles was able to free Prometheus.

Especially since the Romantic era, Prometheus has come to embody a kind of ideal human being. He epitomizes the creative rebel who faced danger, crossed boundaries, and brought progress to humanity through knowledge and technology. Because of his courageous and technological nature, more recently Prometheus has evolved into a proper icon of the enhancement debate, being referred to rather often, even by opponents.[6] But in the current pro-enhancement literature, does he still represent an exemplary yet unattainable idol? Interestingly, in most cases he is actually not presented as the ideal but as the typical human, and, as will become clear below, even as the *true* human.

To return to Stock, he argues that we will not be able to stop human enhancement because we already are like Prometheus: facing dangers, being rebellious, and – since in Plato's version of the myth the Titan is even described as taking part in the creation of mankind – participating in human creation. What is more, it would be unhuman not to steal fire: 'To forego the powerful [enhancement] technologies [...] would be as *out of character for humanity* as it would be to use them without concern for the dangers they pose' (Ibid., my emphasis).

The legal philosopher Ronald Dworkin claims something very similar, urging us not to dread human genetic engineering but rather to dare to self-create:

> Playing God is indeed playing with fire. But *that is what we mortals have done since Prometheus,* the patron saint of dangerous discoveries. We play with fire and take the consequences, because the alternative is cowardice in the face of the unknown.
>
> (Dworkin 2000, p. 446; my emphasis)

Entrepreneur in biotechnology Donrich Jordaan refers to this same quote by Dworkin in an article which criticizes Francis Fukuyama, a political philosopher who passionately argues against biotechnology and enhancement. Jordaan claims that it is precisely Promethean courage that Fukuyama lacks. According to Jordaan, it is thanks to this courage that we live in a modern society today; previously, all was misery, ignorance, and disaster, and we owe it to this courage that we have made such 'awesome improvements to the human condition' (Jordaan 2009, p. 590). And so he concludes his article with the following: 'Beware the day when we betray our *promethean heritage*. Beware the *antipromethean heresy* of Fukuyama' (Ibid., my emphasis).

## Analysis

It seems then that the Promethean ideal, far from being a futuristic image, has rather become the new norm for pro-enhancement theorists. What was first an ideal, unattainable by definition, has now become the standard for what is particularly human. Each of the three thinkers referenced above – let us call them the pro-Prometheans – tries to tell us in his own specific way that essentially, deep down, we are all like Prometheus: creators and rebels, who play with fire and cross natural and technological boundaries. Promethean courage is an indispensable part of our nature, our 'heritage', our cultural constitution.[7] Moreover, they emphasize that we have 'always been' like Prometheus: just like Harris, they claim that we have always enhanced our lives in order to improve our situation; whether literally, by the invention of fire or vaccinations, or otherwise. They want to show that enhancement – in the broad sense – is not as unnatural as many opponents claim.

What is important in this view of human beings as essentially Promethean, and of enhancement as being natural, is that it is not simply a description of what we are, it is not a neutral observation; rather, it implies a strong moral imperative. The claim entails that our *true* nature is Promethean for it is 'too characteristically human' (Stock) to play with fire. It would even be 'out of character for humanity' (Stock) or a 'betrayal' (Jordaan) of our Promethean nature not to engage in enhancement. Thus what the pro-Prometheans are really saying is that we *should* dare to play with fire, *should* participate in our own self-creation, and *should* show this Promethean courage: a true human being enhances.

The question then becomes: but if the enhancing or enhanced Promethean being is the norm, what does that make people who do not use enhancement technologies? Abnormal, perhaps? Dworkin and Jordaan would say cowards. But at a closer look, the unenhanced human may even be seen as dysfunctional; or worse, *unhuman*, as Stock suggests. If we need to show Promethean courage to be who we really are, if enhancement is what makes us 'truly' human, the non-enhancing or the non-enhanced human is a *subhuman* being.

To return to Harris for a moment, his argument is slightly different. He does not literally appeal to a concept of the 'true human', and would probably never want to. However, because of his emphasis on 'the fact that enhancement has always been a [...] part of human development' (2007, p. 4), he does still mean to show that enhancing is something we already do. Moreover, even without this latter claim, Harris' arguments come down to establishing the enhanced human as the norm, since, as stated above, what is most important for him when it comes to evaluating our (health) condition are the possibilities we have. The continuously changing opportunities and possibilities, however, do not determine an ideal and unattainable being, but a norm we should live up to in order not to find ourselves in a harmed condition. If the formerly ideal, enhanced human becomes the new norm, then what was previously normal suddenly degrades to a subnormal state; that is, the non-enhanced human turns into a subhuman being, just as in the pro-Promethean argument. In sum, both Harris and the pro-Prometheans consider the non-enhanced human as suffering from some sort of deficiency.

As an icon of the enhancement debate, Prometheus embodies both the enhanced and the true human. This means that the ancient god – once the romantic, ideal image of the human – descends; he does not become disabled, but neither does he maintain his former perfection. Epitomizing the enhanced human, Prometheus descends from an ideal to a normal status, dragging us mortals along on his descent and thereby forcing us to move over and dismount from our former position. But what are the consequences of this?

To start with the account of the pro-Prometheans, one of the striking consequences is that compared to the standard of the enhanced human, all contemporary humans are in fact subhuman. The above quotes regarding Prometheus show that the human being has shifted from a 'normal' (as in 'common') member of the species, who is striving to become as great as he can be, into a dysfunctional creature as long as he does not enhance himself. In being unenhanced he is unhuman, though he has the possibility of becoming a *true* human being by means of enhancement. What used to be a utopian image, an ideal, has thus become a real possibility. But at the same time, it has made the gap between the creature we actually are and the enhanced, true human we should be wider than ever before, because compared to this true human, the pro-Prometheans consider us to be defective, deficient, or unhuman. And although they do not explicitly say that enhancement is a moral imperative, the Promethean appeal has a clear normative significance.

Once again, as stated above, Harris does explicitly argue that it is our duty to enhance, if possible. In one of his articles (2005) he argues that we also have a moral obligation to conduct biomedical research – to him it seems to be only a question of time until we will be able to radically enhance ourselves. In fact, since age-related research is already looking into the prolongation of life, perhaps we could even say that, for Harris, the theoretical possibility of life-extension also exists as well. This would mean that, according to his standards, in terms of technology we already have radical enhancement possibilities and thus a new norm to live up to. This would imply that in addition to being subhuman, all contemporary humans are also *disabled* as long as we do not do everything possible to develop these and other techniques in order to enhance ourselves. What is certain is that we are disabled in comparison to future enhanced humans.

In other words, bringing these two strains of argument together, we realize that Harris and the pro-Prometheans, by remodeling an ideal image into a norm, end up leaving the non-enhanced human behind as an abnormal, subhuman, disabled, or even masochistic creature – despite these theorists' supposed care for humans.

## Social relations, resources, civil liberty, and the question of the good life

It is, of course, quite an extreme conclusion to assert that all current, non-enhanced human beings are disabled. Nevertheless, it raises important questions, not only about disability *per se*, but also about the implications of this stance with regard to social relations, legal rights, and the effect on individuals. What will be the influence of these new classifications on social relations, for instance? If we are now all disabled, what does that make those who were formerly characterized as disabled? Are they now 'disabled to the second degree'? This does not sound like a particularly neutral phrase, and one would not expect disabled people (according to current standards) to compliantly accept this new label. Moreover, as said, it is virtually impossible to employ such concepts without including some concept of normalcy, norm, or normativity.

Furthermore, the chances are very small that contemporary healthy humans ('healthy' according to the current, commonsensical classification) will suddenly consider themselves to be exactly the same as either disabled people or posthumans. With respect to the latter, bioconservative George Annas claims that it is very unlikely indeed that the contemporary human and posthumans will see each other as equals. 'Instead, it is most likely [...] that we [current humans] will see them [posthumans] as a threat to us, and thus seek to imprison or simply kill them before they kill us' (2001).[8] Alternatively, Annas projects that the posthumans will come to enslave or slaughter us, the 'inferior subspecies' – a 'genetic genocide' could even await us (Ibid.).

Annas' description obviously paints a rather dystopian image of the future. However, it is nevertheless useful to think about the ways in which these new classifications will influence the relationships between groups and individuals in society, including the possible creation of social tensions. It will take time and effort to either change or get used to the new labels for both currently disabled and 'healthy' humans. One source of conflict could be the distribution of resources. If we are all by definition disabled, an ill person – in the traditional sense – does not seem to have any more right to be helped than the 'healthy' do. Perhaps those who are 'disabled to the second degree' would have some priority over others, but under these classifications the difference seems to be smaller than before. Even if this situation has not yet come to pass, and we are not yet all 'disabled', according to Harris we do still require a large amount of resources for scientific research into enhancement. Disputes could arise when significant funding resources are dedicated to every new enhancement possibility, while some people, those currently disabled, for instance, may consider themselves to be more in need of the money – needing it for non-enhancement-related purposes. It would be reasonable for them to refuse to accept the new standards and demand that the funds be dedicated to them, to non-enhancement.

Another question that should be asked is what will happen to civil liberties and legal rights in such a future scenario? Bioethicist Daniel Wikler, for example, has argued that if a majority of people became, by means of cognitive enhancement, vastly smarter than we are now, they would be 'within their rights to deprive the rest of us of our rights, presumably with humanitarian intent' (2009, p. 354). Just as we now 'restrict the freedom of people who are judged to be insufficiently intelligent to handle their own affairs' (Ibid., p. 345), the cognitively enhanced would be entitled to exclude the unenhanced from exercising their civil liberties or equal citizenship.

Although there are no extremes in terms of cognitive difference today, through enhancement technologies a majority of cognitively enhanced people could arise that would be convinced that they knew what was best for us – the current unenhanced humans. In contemporary society, 'healthy' humans do indeed consider themselves authorized to make quite fundamental decisions for people with low intelligence or other cognitive impairments. Some are placed in support homes or mental institutions. In many countries they do not have the right to vote, or if they do, the situation is such that in practice they simply cannot exercise this right. If we compare the current situation to a future one, in which there would be two kinds of people – contemporary and enhanced – it is very likely that, analogous to our current standards and ways of reasoning, the posthuman beings would indeed consider themselves entitled to 'deprive the rest of us of our rights' (as per Wikler's argument above). Moreover, perhaps they could even force those of us who refuse to use enhancement technologies – for either ourselves or others – to fulfill our moral duty so that we might become 'true' or 'better' humans.

A third question one could ask is, what would be the effect on individuals of having a moral duty to enhance? Would this really, as Harris claims, result in 'better people'? Imagine that it is technologically possible to enhance ourselves, and that we are morally obliged to do so. Although it is not entirely clear whether Harris is speaking of a personal duty or a governmental obligation executed through health policy, nevertheless in both situations there would be social pressure to enhance ourselves so that we would no longer be disabled. We would be compelled by society to get out of our rationally identified 'harmed condition' and join the enhanced members of the population, who would already be enjoying many 'worthwhile experiences' that we would be missing. In such a situation we would have two options: enhance ourselves or not. Either way, due to our status as disabled persons with respect to the enhancement possibilities, the very existence of these possibilities could work out quite negatively. Although they promise us a 'better', more 'worthwhile', and 'pleasurable' life, at the same time they establish our condition as disabled and emphasize that we have fewer abilities and capacities than our fellow citizens, and that we are missing supposedly crucial things that would enable us to lead a 'good' life. If I do not enhance myself, apart from the fact that I am not fulfilling my obligation to do so, the sense of duty and the enhancing possibilities that allot me my disabled status will continually confront me with the fact that I am not good enough.

In fairness to Harris, he does acknowledge that the fact that some experiences are closed to disabled people does not mean that they 'cannot find other and different worthwhile things to do and to experience' (2000, p. 98). However, even though I – as a 'healthy' human by contemporary standards, yet a 'disabled' human next to the enhanced – would also have other worthwhile experiences, compared to others and to my own possible alternative conditions I would still be deemed deficient. According to Harris' definition, a rational person – which in practice might very well come down to the majority of the community – would want to be without my disabled condition. Therefore, even though personally I might be satisfied with my condition, on a daily basis I would still have to deal with my difference, with the fact that I lack something in comparison to others. Although I am not bothered by my condition in principle, this situation could still trigger insecurity or dissatisfaction.

If I were to enhance myself, I would be able to enjoy 'the sources of satisfaction' (Harris 2000, p. 98) that were previously closed to me. However, as soon as new possibilities arose, I would fall back into my disabled state and need to improve myself again, leading to the same situation as before. In short, especially when disabled, but even when enhanced, a duty such as the one Harris proposes could very well create a profound feeling of dissatisfaction with ourselves. Even in our 'normal' Promethean state, we would never be good enough for long, but would always see the vulture returning, disabling us again. We would always remain in need of somebody or something else to enhance us further. This would, of course, be a very unpleasant, frustrating, and even alienating feeling. Perhaps, instead of encouraging a person to continually improve himself, such a duty would have the reverse effect; that is, *hamper* a person's aspirations and drive for action because he would perceive his aims as unachievable relative to his means, skills, and capacities. The question, therefore, is whether a moral duty to enhance would not actually accomplish the opposite of what it is intended to do, and create, instead of 'better people', unsatisfied, frustrated, and self-alienated beings, awaiting their Heraclean salvation.

## Conclusion

The Promethean drive to improve, transform, and create thus might end up not being as positive as expected by those in the pro-enhancement camp. I do not mean to say that the adventurous urge is without value, or that we should dismiss all Promethean enthusiasm, as some opponents do. However, I think that in realizing the difficulty of establishing a kind of being or behavior as normal, we should not immediately dissolve the line between therapy and enhancement, but rather acknowledge how difficult it is to classify humanity at all without implying some concept of normalcy and its corresponding notions of normativity. As we have seen, Harris himself implicitly introduces something that was formerly an ideal – the enhanced human – as the new norm. And despite his rejection of normalcy, in his case too what counts as disabled depends on both physiological and relational – social – factors, and is determined relative to the norm. The concept of normalcy will probably always be present to a certain extent, whether we like it or not. Therefore, I think it is of great importance to be aware of and recognize the use of the concept of normalcy and its associated values, its influence on the concept of disability, the position it allots to enhancement, and, most of all, its impact on society and (groups of) individuals when classifications change.

## Notes

1. As formulated by C. Boorse (1975), and used by many others.
2. The concept of normalcy is not neutral and unproblematic, but rather loaded with values and notions of normativity that vary according to context (see Scully and Rehmann-Sutter 2001). For instance, it establishes disability as 'abnormal' since it is defined by reference to its 'departure' from normal functioning. I will address the issue of normalcy later.
3. For a bioethical discussion of the issue, see Scully, J. L. (2008) *Disability Bioethics: Moral Bodies, Moral Difference*. Lanham: Rowman & Littlefield.
4. For a discussion of 'Deaf culture', see Blume, S. (2010) *The Artificial Ear: Cochlear Implants and the Culture of Deafness*. New Brunswick, New Jersey: Rutgers University Press.
5. There are many pro-enhancement theorists who argue along similar lines – Nick Bostrom, Gregory Stock, Donrich Jordaan, Kevin Warwick – some of whom I will discuss in this chapter. Furthermore, feminist and postmodernist Donna Haraway's claim that we are all 'cyborgs' – 'theorized and fabricated hybrids of machine and organism' (1991, p. 150) – could also arguably be interpreted as meaning that we are all enhanced.
6. Of course, opponents refer to Prometheus precisely to warn of the risks of enhancement, emphasizing (though often implicitly) the negative sides of the myth.
7. Jordaan, by labeling Fukuyama guilty of 'antipromethean *heresy*' (my emphasis), even insinuates that the notion of Promethean courage has some religious sanctity.
8. See http://www.gjga.org/inside.asp?action=item&source=documents&id=19&detail=print (last visited 23 March 2012).

## References

Anderson, W. F. (1990): Genetics and human malleability. In: *The Hastings Center Report*, 20(1), pp. 21–24.

Anderson, W. F. (1994): Genetic engineering and our humanness. In: *Human Gene Therapy*, 5(6), pp. 755–760.

Annas, G. (2001): Genism, racism, and the prospect of genetic genocide. presented at the UN World Conference Against Racism, Durban, South Africa, 3 September 2001, http://www.gjga.org/inside.asp?action=item&source=documents&id=19&detail=print, date accessed 23 March 2012.

Blume, S. (2010): *The Artificial Ear: Cochlear Implants and the Culture of Deafness.* New Brunswick: Rutgers University Press.

Boorse, C. (1975): On the distinction between disease and illness. In: *Philosophy of Public Affairs,* 5, pp. 49–68.

Bostrom, N. (2008): Why I want to be a posthuman when I grow up. In: Gordijn, B. and Chadwick, R. (eds.) *Medical Enhancement and Posthumanity.* Dordrecht: Springer.

Daniels, N. (2001): Justice, health and health care. In: *American Journal of Bioethics,* 1(2), pp. 3–15.

Daniels, N. (2009): Can anyone really be talking about ethically modifying human nature?. In: Bostrom, N. and Savulescu, J. (eds.) *Human Enhancement.* Oxford: Oxford University Press.

Dupré, J. (1998): Normal people. In: *Social Research,* 65(2), pp. 221–248.

Dworkin, R. (2000): *Sovereign Virtue.* Cambridge: Harvard University Press.

Haraway, D. (1991): A cyborg manifesto: Science, technology, and socialist-feminism in the late twentieth century. In: *Simians, Cyborgs and Women: The Reinvention of Nature.* New York: Routledge, pp. 149–183.

Harris, J. (1993): Is gene therapy a form of eugenics? In: *Bioethics,* 7(2/3), pp. 178–187.

Harris, J. (2000): Is there a coherent social conception of disability? In: *Journal of Medical Ethics,* 26, pp. 95–100.

Harris, J. (2005): Scientific research is a moral duty. In: *Journal of Medical Ethics,* 31(4), pp. 242–248.

Harris, J. (2007): *Enhancing Evolution.* Oxfordshire: Princeton University Press.

Jordaan, D. (2009): Antipromethean fallacies: A critique of Fukuyama's bioethics. In: *Biotechnology Law Report,* 28(5), pp. 577–590.

Newell, C. (1999): The social nature of disability, disease and genetics: A response to Gillam, Persson, Holtug, Draper and Chadwick. In: *Journal of Medical Ethics,* 25, pp. 172–175.

Reindal, S. M. (2000): Disability, gene therapy and eugenics: A challenge to John Harris. In: *Journal of Medical Ethics,* 26, pp. 89–94.

Scully, J. L. (2008): *Disability Bioethics: Moral Bodies, Moral Difference.* Lanham: Rowman & Littlefield.

Scully, J. L. and Rehmann-Sutter, C. (2001): When norms normalize. The case of genetic enhancement. In: *Human Gene Therapy,* 12, pp. 87–95.

Stock, G. (2003): *Redesigning Humans.* London: Profile Books.

Wikler, D. (2009): Paternalism in the age of cognitive enhancement: Do civil liberties presuppose roughly equal mental ability?. In: Bostrom, N. and Savulescu, J. (eds.) *Human Enhancement.* Oxford: Oxford University Press.

# 10

# More Human than Human![1]: How Recent Hollywood Films Depict Enhancement Technologies – And Why

*Kathrin Klohs*

## Introduction

> Ladies and gentlemen, welcome to the next generation of science: The agnate. An organic frame engineered directly into adulthood to match the client's age. [...] Within twelve months, it will be harvest-ready, providing a carrier for your baby, a second pair of lungs, fresh skin...all genetically indistinguishable from your own. [...] It's a product, ladies and gentlemen, in every way that matters.
>
> (*The Island* 0:49)

This warm welcome, from Michael Bay's movie *The Island*, is a mad scientist's advertisement designed to sell the organs of artificial human beings. By breaking the law against human cloning, this character fulfills the financial interests of his huge company, creates for himself the opportunity to play God, and makes the new American Dream of eternal life come true – for the rich. Examples from science fiction like this one are frequently used in science: as introductions, illustrations, 'eye catchers', or just as jokes. Novels and short stories, movies and comics are used to attract the reader's or listener's attention in academic writing, lecturing, and teaching, and as an effective device for provoking collegiate discussion. But is that really all they are good for?

Assuming that fiction is able to contribute to public debate about bioethics as well as reflect it, the following remarks concentrate on human enhancement in recent movies. In this context, 'enhancement' is understood as any 'betterment' of the human condition that goes beyond common medical interventions:[2] by manipulating the genome of the species, by altering the genotype or phenotype of an individual, or by transferring consciousness from one receptacle to another. This paper will first of all sketch the main traditions and aesthetic strategies of representing science on the screen. The movies *Gattaca* (1997, directed by Andrew Niccol), the *Matrix* trilogy (1999 and 2003, directed by the Wachowski brothers), *Equilibrium* (2002, directed by Kurt Wimmer), *The Island* (2005, directed by Michael Bay), *Surrogates* (2009, directed by Jonathan Mostow), and *Avatar* (2009, directed by James Cameron) are then examined in order to highlight three points. First: what is human nature according to the fictitious societies depicted in these films, societies which have become familiar with human enhancement? Second: what kind of society will – or could or should – evolve due to human enhancement, according to these fictitious societies in which it has been legalized? Third: will – or should – everything we are capable of doing be done, according to these fictitious societies in which most of it has been done?

To reach these goals, I will not go through moral or ethical arguments as would a bioethicist, explain current or future healthcare possibilities as would a physician, nor will I give a report on practical research as would a geneticist. Instead, I will talk about philosophy, society, and film, in particular about these specific movies' underlying definitions of being human, and the narrative and artistic strategies that make the audience imagine a posthuman future by both indicating and co-constructing the enhancement discourse. The forms of enhancement that these movies offer as a tool, and the societies they present as a result will launch a discussion about the human condition, the rights and obligations of the individual, equal opportunities and distributional justice, freedom of choice and freedom of action, and sustainability and risk perception.

## I'm a cyborg, but that's ok:[3] Bioethics at the movies

First, we must ask why it is relevant to discuss human enhancement using examples from film at all. Alterations to and spectacular transformations of the body, presented as either 'good' and necessary ('therapy'), or as 'bad' and ambitious ('enhancement'), have always been topics for science fiction. Genetic engineering and reproductive medicine in particular have featured in this genre since the 1970s (cf. Stacey 2010; Wulff 2008). But why exactly is it worth discussing fictional representations of human enhancement within a medical and ethical framework? What is the epistemic potential of cultural images in such films? Can bioethics gain or adapt something from the possible worlds represented in fiction? Does the audience learn from depicted realizations of thought experiments? And how should the scholar read movies about bioethics? Obviously, the interaction between fiction and reality must not be underestimated: not only are science fiction films inspired by current research, but they themselves inspire scientists, technicians, and engineers, and change the way lay people think about the future. Alfred Nordmann has even suggested the creation of a research committee in which scientists are paid to read sci-fi books and pick out ideas worth funding (Nordmann 2010). Mike Michael and Simon Carter argue

> If [. . . ] fictional genres enable particular readings of science, scientific knowledge is, conversely, necessary for the reading of certain of these fictional genres. What we have then is an interesting twist to the relations between science and public. Scientific knowledge is necessary to 'read' these fictional texts (in the broadest sense), yet the 'interpretation' and the discourses, metaphors and narratives of fictional genres may shape 'understanding' of science.
>
> (2001, pp. 11–12; cf. also Shapshay 2009)

Below I will sketch out the central arguments of what films can contribute to a discussion of human enhancement, though without going into scholarly debates in the research fields of narratology or film studies.[4] First of all, in the arts and humanities, storytelling is regarded as an innate drive, a part of human nature that strives to make sense of life. People conceptualize their lives as stories and themselves as the protagonists and, for the most part, are affected far more by stories they are told than by logical argument and rational reflection. A movie about bioethical issues is completely different from a bioethical debate due to its narrative situation: it tells a single, exemplary, and concrete story. The pros and cons of an academic dispute or a public argument come to life and turn into a conflict that initiates action and requires characters. Thought experiments must leave behind abstraction and surrender to consequences and endings. Enhancement visions, in particular, may still lack real-life case studies, so that examples can only be derived from fiction, to say nothing of the fact that, for most people, issues that do not form a part of their everyday experience are only accessible through mass media (cf. Maio 2008, p. 291).

Different cinematic strategies also help to present the enhancement topic. Philosophical issues are shown in an exciting, often fascinating way: 'Science is true but boring; SF [science fiction] is less true but exciting. [...] If science fiction was deficient in facticity, science fact was deficient emotionally or motivationally' (Michael and Carter 2001, p. 18). It has been well known from the beginning of film history that camera shots mainly work physiologically, and therefore have an affective impact on the viewer who identifies with what is shown long before intellectual processes emerge. The combinations and resulting effects of different visual and acoustic elements – such as music and sounds, colors and light, patterns and shapes, field size and point-of-view shots – directly speak to the senses and depict the characters and emotions. In addition to this stylization, scriptwriters and directors refer to the basic forms of narration and add the characteristic elements of thrillers or love stories in order to entertain spectators of all kinds. For the same purpose, Hollywood stars are cast, 'who themselves might be the products of a program of genetic perfectionism' (Gavaghan 2009, p. 75).[5] Concerning the set, it is customary to overplay the beauty of the brave new worlds, especially in contrast to their apocalyptic surroundings.

Moving on to the epistemic level, it can at least be claimed that a work of art articulates both the artist's own influence and various collective attitudes, dreams, and fears, especially in popular culture. When dealing with human enhancement, the movies thus depict the ambivalence of Western societies concerning progress and science; technophile as well as technophobe visions are omnipresent.[6] The audience is free to identify with a creative and omnipotent inventor; with an enhanced being that is able to transgress the borders of the possible in terms of body and mind, time and space, consciousness and self; or with a totally powerless victim under the controlling eye of Big Brother medicine, who is subject to painful procedures. To go even further, movies more or less explicitly cite myths and *grands récits* – be they archetypes such as Prometheus, whom Ronald Dworkin has called the patron saint of dangerous discoveries (2000, p. 446),[7] or modern attitudinal phenomena like the American Dream. Thus, the audience is led to identification and interest, tension and fear, pleasure and disgust.

Accordingly, Sandra Shapshay argued that 'films can get us to see the world differently [...] by engaging us in a cinematic world, emotionally and intellectually' (2009, p. 2). Having focused upon the former, the possibilities of the latter must also be reconsidered. First of all, for transhumanist and posthumanist thinking, fiction has always been an opportunity to think through speculation and bring it to some kind of conclusion.[8] Asking 'what if...?', fiction is a sheltered place in which to sketch tomorrow's wanted and unwanted consequences of today's ethical decisions. Continuing an initial question from bioethics, emplotment also enables filmmakers to add some new – maybe underestimated – perspectives: movies paradoxically discuss a scientific topic from outside the scientific community, and manage to take into account the views of a lay audience. Given that in fiction we can distinguish narration itself from the story, expert knowledge may be represented by scientists (acting in an advisory capacity at production level or as characters at story level); the lived experience of those affected may be introduced through 'enhanced' as well as 'non-enhanced' characters (in their depictions, dynamics, and constellations); and tacit knowledge may be transported implicitly by narrative possibilities (such as focalization, reliability, or comment) and cinematic strategies (like music and sound effects, camera position/movement/angle, or *mise en scène*). Different kinds of knowledge, including non-knowledge, are thus represented on the screen, competing and overlapping. Finally, movies about human enhancement hold the potential to disturb our image of the human body, to show and to establish a radically new notion of beauty. They can show altered beings not as a shocking problem, but as an alternative to the 'normal'.

## Here I am. Doing science:[9] Science at the movies

Having shown why it is important to look at fictional representations of bioethics, this section will briefly mention some traditional means of expression in film history and summarize the movies investigated. When it comes to depicting science in a piece of art, contemplation and thought in the humanities are the obvious topics to reflect upon in written texts (due to the various possibilities of presenting the content of characters' consciousness; cf. the campus novel genre), whereas experiments and discoveries in the 'hard sciences' are more commonly represented on the screen (due to the visibility of the results that matches a visually based medium).[10] Thus, certain patterns of presenting scientific work and characterizing scientists emerged at the very beginning of film history and have remained remarkably unchanged; standardized appearances and objects have become common practice (see Frayling 2005; Perkowitz 2007; Weingart 2005).

Fritz Lang's *Metropolis* (1927) provides an early and paradigmatic illustration of this. In the movie, the audience meets Rotwang, an inventor (Rudolf Klein-Rogge), who tries and ultimately succeeds in creating an artificial being due to some form of uploading. The movie displays some of the features that are still crucial to the cinematic depiction of science, in particular the stock character of the 'mad scientist', a white male who is a genius in the natural sciences but who is also evil, crazy, dangerous, and eager to rule the world – through finding the philosopher's stone, the fountain of youth, or the elixir of life. He may be characterized as megalomaniac and obsessive or as absent-minded; his experimental work may take place in a secret dungeon or a research center maintained by the government. Fritz Lang also uses a stock setting and stock props: a laboratory that consists of wires and cables, measuring instruments and display boards, test tubes and bubbling liquids. Chemistry and physics are invoked, man–machine links are offered. We encounter similar set-ups in today's science fiction films, and although the equipment has shifted from mechanic to virtual, and the scholar's books have disappeared to be replaced by touchscreens, the cinematic strategies have hardly changed.

The transhumanist and posthumanist visions in the movies introduced below are derived from this cinematic tradition. They all deal with various aspects of human enhancement, which is less the restoration of health than the betterment of the body, the brain, the emotions, or consciousness.

As far as the debate on bioethics is concerned, most film reviews and research on film (including Clayton 2003; Easton 2008; Gavaghan 2009; Kirby 1999; Matrix 2006; Platzgummer 2003) are aware of Andrew Niccol's *Gattaca*. The movie is about a boy who was conceived without his parents consulting a geneticist, in a future world of 'genoism' based on reproductive technologies, where education and employment, insurance and income, social standing and marital status depend on one's genetic information. A new kind of two-tier society differentiates the enhanced from the non-enhanced, implicitly treating the non-enhanced as disabled and deficient. The protagonist's (Ethan Hawke) passionate aspirations for a career in space thus make cheating the system inevitable. He does a deal with a member of the elite, Jerome (Jude Law), and uses this man's identity while his own mind and personality succeed in his career and love life. Accordingly, his character sums up ideas of justice, freedom, and self-determination.

As previously mentioned, in *The Island*, a mad scientist illegally undertakes breeding on a mass scale and then kills his clones in order to sell their organs as spare parts to wealthy clients in need of enhancement. But the designers have underestimated human curiosity; the clones ask inappropriate questions, eager to know about the past and the future, the why and what for. When the protagonist (Ewan McGregor) and his girlfriend (Scarlett Johansson) escape and discover the real world, the company decides that hundreds of clones must be killed and redesigned to prevent a repetition of such an event. The audience witnesses their dramatic rescue.

In *Equilibrium*, after a life-threatening Third World War, the government has concluded that all cruelty and violence results not only from pain, jealousy, and anger, but also from love, admiration, and desire. Thus, medication is forced on everyone to enhance people's inner state and to prevent the human ability to feel. People are balanced but suppressed and stabilized using a psychiatric drug called Prozium, which constitutes the main topic of omnipresent propaganda. The viewer is informed about a leading functionary (Christian Bale) who breaks with the system, secretly refuses his medication, and begins to explore human emotions while still in charge.

Leaving behind these examples of bodily based modification and therefore transhumanist predictions, I will now give three examples of posthumanist visions in film, all of which feature virtual reality and mind uploading or mind transfer.

In *Surrogates*, enhancement is pushed forward by extending the human brain via an external object. A new kind of robot is invented with the consequence that 98 percent of the world population does not even dress in the morning, but just plugs in. Sidestepping their bodies, their brains directly control machines that transfer sensory perception and act out their daily routine. But the price to pay is a diminishing of life: the main character (Bruce Willis) hates his out-of-body life and desires true human experience, which means discovering the world with all one's senses. Accordingly, he and his wife (Rosamund Pike) learn to appreciate their imperfect, vulnerable bodies more than those of their flawless robots.

Avatars, bodies one imagines inhabiting when playing a computer game, are the main characters in James Cameron's movie of the same name. A war veteran (Sam Worthington) makes contact with aliens in a scientific experiment and recognizes that they show up with a fascinating mixture of nature and technology. He is enhanced whenever his consciousness is transferred from his invalid human body to the avatar that is a quasi-enhanced amalgamation of human and superior alien DNA. Thus, mind uploading means to overcome the limitations not only of his paraplegia but also of the human condition.

Finally, the *Matrix* trilogy features a war between humans and machines. The world we know is presented as an illusion maintained by the latter and forced upon the former. People think they live their lives but they are only in a dream, a reference to a famous philosophical thought experiment: 'How does it feel to be a brain in a vat?' (discussed, for instance, in Putnam 1981, ch. 1). In fact, humans produce energy, harvested by machines that have turned against their inventors. Artificial intelligence (AI) has developed its own interests and humankind is powerless to stop its ultimate rationality. The story is about a young computer hacker (Keanu Reeves) who is expected to end the war and save the human race.

## What do you think I am? Human?[11] What is suggested as human nature?

Implicitly or explicitly, all of these movies present certain qualities as inherent to our species. As can be seen above, their stories and characters rely on key properties that suggest what is – and always has been typically human. There is a wide array of individuality, freedom of choice, and self-determination; ambition, goal orientation, and fighting spirit; curiosity, epistemology, and quests for the purpose of life; emotionality, affectivity, and sensorial nature; imperfection, vulnerability, and mortality. On the screen, today's humankind is faced with tomorrow's possible side effects of AI and enhancement technologies. Today's non-enhanced beings are abandoned in or fascinated by future societies that take human enhancement for granted. It is not the norms and definitions of human nature that have diachronically changed – it is only the mirror foils. Instead of contrasting humans with angels or gods, science fiction uses robots, cyborgs, or aliens for the same purpose. Countless details still refer to the Western tradition – for example, the company's clones in *The Island*, that threaten the clients' uniqueness, reference the Romantic tradition of the doppelganger (*revenant*) and function as heralds of death within the story (cf. Wulff 2008, pp. 273, 281). To provide a deeper understanding of this tradition, a digression on the *Matrix* trilogy is appropriate.

## Digression: Mirror foils of the technological age

In its second part, *Matrix Reloaded*, the main character meets The Architect, a personalized computer program. It is a brilliant piece of directing and stylization to present the situation on the screen in such congruence with the human idea of God. Not only is the appearance of the old man with white clothes and a white beard an allusion, but he also forces the protagonist, who represents the human race, to make a decision between good and evil. Additionally, his omniscience is visualized by a shining spherical room covered with television screens on which the viewer can see all the historical stages of humankind simultaneously. It is common practice to use visual patterns such as this to present any creation with the aid of analogy: creatures are to creators as humankind is to God. (This is also illustrated, for example, by the relationship of writer and character in *Stranger Than Fiction* [2006, directed by Marc Forster]; the relationship of programmer and computer game in *TRON: Legacy* [2011, directed by Joseph Kosinski]; the relationship of inventor and robot in *Blade Runner* [1982, directed by Ridley Scott].) But the movie goes further inversely: whereas the human intellect is traditionally conceptualized as inferior in contrast to the presumed perfection of God as its creator, the creator is now transcended by AI, its own creation, while the figure of thought and the strategies of visualization remain the same. The protagonist is told that the matrix consists of harmonic, mathematically precise sequences, although these surroundings are incompatible with its residents. Consequently, what happens in the matrix due to human nature is beyond The Architect's understanding as it would require 'a lesser mind or perhaps a mind less bound by the parameters of perfection' (*Matrix Reloaded,* 1:49). Constructed by human beings, the machines themselves have developed programs that can no longer catch any meaning in the human way of thinking. This addresses a crucial problem of AI as well as of religious faith: in some measure, intelligence may be regarded as gradually different, but at a certain point, a qualitative change takes place and it is neither downwardly nor upwardly compatible any more.

On the other hand, there is another personalized program – Agent Smith (Hugo Weaving), the antagonist – that alludes to the human concept of the devil, and aggressively discusses human mortality, the fragility of the human body, and the meaning of life in *Matrix Revolutions,* the trilogy's final part. Smith temporarily manages to inhabit a human body and describes his new experience: 'it is difficult to even think encased in this rotting piece of meat. [...] Disgusting. Look at how pathetically fragile it is. Nothing this weak is meant to survive' (*Matrix Revolutions*, 0:51).[12] AI considers itself a more highly developed, stronger species that will surpass the human race. Leaving behind the finite nature of life, pure thinking claims to be the next step in a process that is no longer defined as the evolution of humankind, but as the evolution of intelligence, and is therefore no longer told from a human point of view. Accordingly, when the survival of his species is at stake, the digital antagonist Smith tries to discourage the human protagonist Neo, arguing in a nihilistic way:

> Why keep fighting? Do you believe you're fighting for something? For more than your survival? Can you tell me what it is? Do you even know? Is it freedom or truth? Perhaps peace? Could it be for love? Illusions [...]. Vagaries of perception. Temporary constructions of a feeble human intellect trying desperately to justify an existence that is without meaning or purpose.
>
> (*Matrix Revolutions*, 1:44)

In the cases of both The Architect and Agent Smith, being human is defined in comparison to ideas that have been well known for generations but without regard to the vehement challenges that genetic engineering poses to any essentialist definition. In this regard, the movies discussed in this paper fall short of their emancipatory potential, as none of them show an unheard of concept of human nature or overcome the dichotomy between the well known and the unknown, the normal and the abnormal, the healthy and the diseased – which would mean to propose diversity or to call into question the assumption that mortality, sickness, or disability are self-evident misfortunes in need of cure. Other pieces of art (such as Marc Quinn's sculpture *Alison Lapper Pregnant*, 2005) are much more awakening; and in real life as well, the common sense of an appropriate body restoration is indeed defied (consider the disabled athlete Aimee Mullins, who appears with more than ten disconcerting pairs of artificial limbs for different purposes). In this context it would be interesting to have a closer look at the *X-Men* movies, where mutants fight for integration into a future society and cement their status as their own, third category; or at David Fincher's *The Curious Case of Benjamin Button* (2008), which manages to show some respect for its protagonist (Brad Pitt), who is born elderly and grows young but is never medicalized despite his otherness. An analysis of these movies would also contribute to the analysis of disability in Hollywood movies: according to the cinematic depiction, is cure to disability as enhancement is to health? Do the films use the limitations of the human condition as a metaphor for disability? Or vice versa?

Given this concept of being human, the question arises of whether the concepts of a social system are also heirlooms from the Western tradition. Thus, shifting my focus to social life, I will move on to examine future societies in the movies that already practice different forms of enhancement and exaggerate their consequences: what rights, obligations, and opportunities arise as a social effect? Are possible body modifications presented as harmful or fruitful? Is humankind freed or enslaved?

## No one exceeds his potential:[13] What kind of society will – or could or should – evolve?

Both positive and negative utopias appear in the movies. On the one hand, the endless possibilities for manipulating the world, the infinite choices for altering oneself, and a holistic social structure do have a certain allure. On the other hand, the commercial exploitation of the human genome, a loss of self-determination over body and mind, and a continuation of totalitarian structures all appear as threatening.

To start with the pessimistic view, the Gattacan society (see Bühl 2009 and Gavaghan 2009 about this term) rigorously discriminates between the so-called 'valids' – a new elite whose parents ordered optimum hereditary factors from a geneticist, who gain access to superior employment, and who indulge in a luxurious lifestyle facilitated by their income – and the so-called 'in-valids' – a new underclass, children who were conceived in the natural way, accidentally or because of poverty, who are accused of being a 'utero', 'faith birth', or 'de-gene-rate'. Defined *ex negativo* as not valid and not even normal, these 'in-valids' are discriminated against and have no chance of climbing the social ladder.

> GATTACA presents one possible configuration of the relationship between the coevolution of genetics and global, capitalist, neocolonialist, and corporate interests – adopting a critical, even a dystopic view of the future of genomic technoscience.
>
> (Matrix 2006, p. 102)

Today's audience's categories are blurred. On the one hand, as illustrated by Vincent's character, the 'in-valids' are treated as sick and abnormal in the Gattacan society although they would be looked upon as healthy and normal in our society. Their hereditary factors are regarded as inferior because they were naturally conceived and will therefore never touch the optimum genetics of the 'valids' who were composed *in vitro*. On the other hand, Jerome's character gives an example of a 'valid' who holds an enormous potential but yet falls short of the capability even of the 'in-valids'. He is born genetically superior but *de facto* paraplegic as a result of a failed suicide attempt and thus forms a contrast to both normalcy and disability. The stories of Jerome and Vincent both show how bodies are judged of worth in the future. Paradoxically, norms tell what is 'abled' and 'disabled', and at the same time abilities and disabilities shape the understanding of these norms. This way, otherness and disability become visible primarily as a social, not a medical state.

In *Surrogates*, as well, in contrast to the omnipresence of breathtaking and encouraging commercials, customers who have little money are fobbed off with cheesy, rugged robot models, which are provided with only basic functions and are sold in second-rate shops. In short, in both movies what people are depends on what their parents could afford, and what they can afford for themselves or their children depends on what they are. If they are not genetically optimized, they will fall by the wayside. Be it reproductive and genetic technologies, organic or mechanical spare parts for the body, or a simulated reality for the mind: genes and limbs, thoughts and feelings are sold like anything else in the movies. Whereas in former times the social standing of the ruling classes was established through education and maintained through performance, now their superiority has become obvious, provable, and irreversible. What allegedly 'ran in the blood' is now literally 'em-bodied' on the screen.

But even those who are willing and who can afford to cooperate do not own their enhanced bodies and minds. In these films, most characters lack privacy, forced by their government or their employer to obey rules for optimum health and to put their bodily functions under the control of all-embracing medical supervision. Figuratively, from the point of view of the polity and the labor market, an individual only counts in terms of his or her monetary value as biocapital.[14] In *The Island*, the clones cannot make a hedonistic choice against a healthy life because they are products, designed to be sold in showroom condition. In *Gattaca*, people are constantly forced to provide urine for substance tests and blood for security checks, so that selling bodily fluids, hair, and eyelashes becomes a branch of business in secret networks (watching the periodic close-ups of medical equipment and extraction processes may fill the audience with disgust). It is important to note that even if the white male protagonist in *Gattaca* keeps up the fight by himself, succeeds due to his strong belief, and founds a new heroic myth of progress, the regime in the movie is subversively betrayed only in his particular case, but not overthrown or changed by his actions (unlike in *The Island*, *Equilibrium*, and *Surrogates*). And even the immaculate outcome of genetic engineering feels uncomfortable. Jerome, whose middle name happens to be 'Eugene' (the one with good genes), is born unblemished but feels pressurized, flinches from ambitious and exaggerated expectations, and self-immolates in the end. His optimum genetic make-up has set the stage for overwhelming success in life, but he takes his perfection for a burden.

Additionally, the way in which these totalitarian characteristics are depicted in most cases alludes to fascism, imperialism, colonialism, or the transhumanist concept of 'speciesism' – and therefore serves as a metaphor for well known political issues. This is obvious at the story level in *Avatar* and at the aesthetic level in *Equilibrium*. The latter depicts the restrictions, forces, and constraints of its enhancement-based totalitarian society. Fascism is cited through (swastika-like) symbols, the architecture of the utopian setting (referring to pre- and post-war Berlin), and the excessive use of a leather-clad police force. Most pictures are horizontally striped and develop perfect symmetry so that the audience feels a sense of the timelessness, stability, and gravity of a social order that is not to be changed. Restricted to black and white, and grey and beige, uniformity, functionality, purity, and clarity is shown; patterns, paintings, odors, music, toys – everything that disrupts the senses – is damned and hidden. As the characters exist without emotions due to their drugs, the movie manages to translate their subjective inner life to the visual level, and make it accessible to intersubjective perception.

In spite of these negative utopias, there are also strong positive aspects. It is crucial to the narrative and aesthetic strategies of the movies that the future world in which the characters live opens up radically new perspectives as a result of human enhancement. First, all sorts of alterations can be acted out in the categories of time and space – the surprising, award-winning special effects of the movies depict thrilling adventures and spectacular combats just because both take place in virtual realities or are fought between enhanced bodies that are no longer limited by physical conditions. Second, the characters' options in choosing their own or their offspring's appearance, sex, skin color, character, talents, and even memories can be read as a prolongation of the American Dream that is obviously connected to recent theoretical approaches aimed at leaving behind race and gender. Third, one of the movies, *Avatar*, presents the contrasting exception of a utopian dream worthy of detailed discussion. An alien society serves as a model, mixing our nostalgic dream of going back to our roots with the desired future of information technology (in general, the audience is used to having these presented as opposites because modern science is based on subjugating the 'world'). Mother Nature is worshipped as a personalized goddess on the foreign planet Pandora; male-dominated human science is still in its infancy compared to her wisdom – what a gorgeous family photo. Paradoxically, the planet offers organic access to a data network that is not technically developed but consists of the branches and leaves of a giant tree. Instead of man–machine interfaces, well known from science and science fiction, *Avatar* presents alien-animal interfaces (to tame livestock), alien-tree interfaces (to connect with cultural heritage and the ancestors' wisdom), and alien-planet interfaces (for religious experiences). Consciousness can also be understood as data and can be transferred from one body to another by linking both receptacles to this network. Bodies, as foreseen, are treated as interchangeable containers, temporarily holding a consciousness that is potentially eternal. Network visions – as developed in today's science labs – and concepts of collective consciousness and collective memory – as recently discussed in the humanities – are intermingled to show a sustainable society that is based on the exchange of information.

By looking at their definitions of human and social life, it has become clear so far that the movies take a position regarding whether the implications that enhancement technologies might bring about in the future are desirable or not. But are these predictions on the screen congruent with science (cf. Grebowicz 2007; Michael and Carter 2001; Perkowitz 2007)? Is it even possible to predict such advances? If not, what is the subject of the movies?

## So when did killing become a business for you?[15] Will – or should – everything we are capable of doing be done?

The following section will discuss the distinction between science fiction on the one hand and future science on the other, as well as the distinction between risk and danger. In order to explain the extent to which the cinematic representations of human enhancement are diagnostically conclusive, I will outline three considerations: the momentum of scientific progress, the psychological dimension of enhancement visions, and the limits of human creativity.

In general, the ability to predict advances in the natural sciences, and therefore in enhancement technologies, is severely restricted. First, we have no idea how things will develop when an experiment steps out of the laboratory and into the human body. Second, given the fact that knowledge and technological progress are growing at an ever-increasing rate, and considering effects such as positive feedback, exponential growth, and self-improvement, it is impossible to predict what may happen; a point that is often discussed under the term of singularity. Third, if AI becomes more intelligent than human beings, and if it is able to improve itself or construct even more intelligent machines, we simply cannot imagine what kind of thought will be set free. Instead, life-threatening monsters as well as the *deus ex machina* in movies about human enhancement should be explained differently: science fiction predominantly deals with emotional problems that are psychoanalytically classified as characterizing the children – especially the sons – of the middle classes (e.g., Seeßlen 2000). In a genre about the so-called rational (culture, men, mental health) that splits off and controls the so-called irrational (nature, women, madness), what is suppressed inevitably returns but cannot be recognized. In its place, the protagonists have to overcome monsters and catastrophes – not as real dangers but as unwanted by-products of contradictions in society, based on the rational-scientific worldview. Furthermore, the fact that creatures of all kinds are envisioned in such an anthropomorphous way is rooted in the boundaries of human imagination. As already discussed in Cartesian philosophy (Descartes, *Meditationes De Prima Philosophia* I, 8), inventing or predicting something radically new is in fact impossible. Why, for instance, should a superior intelligence build cities, fight with firearms, or go backwards in terms of evolution to organize its shape by imitating sea creatures like mussels or jellyfish – as the machines in the *Matrix* trilogy do? This is improbable, but only comparison provides an opportunity to imagine a new species in an understandable way. Science fiction finds it hard to create either societal models or terrestrials and aliens out of nothing. The human intellect falls back upon what it knows, playing a game of decombination and recombination, and thus constructing chimeras. But in fact, AI will probably not care about human beings, their goals, values, dreams, or fears. It will not be inherently evil or actively malicious, will not be humankind's best friend nor will it solve all of its problems. Machines will do the things they are constructed to do, whether their actions are connected to our affairs or not.

Nevertheless, there is one thing about human enhancement that we can predict with certainty, and that is clearly shown in all of the movies: instead of considering diseases, disabilities, and death as *danger*, they will be seen increasingly as a *risk*. To differentiate sociologically between these terms (cf. Hijikata 1997; Luhmann 1990; Münkler et al. 2001), people are threatened by a danger, something that they cannot control, avert, or prevent but can only suffer or accept; by contrast, a risk rests on decisions that depend on probabilities, calculations, and approximations people have to act and make choices, which sometimes causes guilt and remorse. Obviously, the more decisions are feasible, the more it becomes a question of risk rather than of danger. Once the next generation's intended physical and mental condition can be defined and ordered, parents who decide to father a child by sexual intercourse *fail* to do all they can for their offspring. It may be that, having acted on their own authority, they have to pay punitive insurance premiums and take their chances trying to access unaffordable medical treatments, since the public purse can no longer be relied upon. As a result, it may be regarded as inexcusable negligence or even as a sex offence to give birth to a child that is conceived by chance. Using the example of *Gattaca*, who will pay for a 'de-gene-rate' boy who in all probability will need expensive therapy for various diseases or will be depressive, aggressive, or unable to learn? The child itself will be judged less by its individual merits and endeavors and more by its genetically determined potentials and probabilities. To stay with the same example, is the protagonist in *Gattaca* really treated in an unfair way when, despite his hard-earned astronautical skills, he is excluded from an expensive space mission due to the risk that he may die from a congenital heart condition? Maybe he is the one who violates ethical principles by betraying the company (cf. Gavaghan, pp. 80–81)? Given that what is predicted from the genetic code is probable but not certain to happen, do people who are not equal deserve equality of opportunity? Science fiction thus heralds a new ethics that may force people to shop in the genetic supermarket instead of playing the genetic lottery. In the end, however, we recognize the possibility of risks turning back into dangers. What was invented to reduce the impact of danger paradoxically gathers momentum and turns out to be the biggest danger of all.

## Conclusion

I will summarize the main points of this paper. At the start, making claims about the strong points of the medium of film, from storytelling to an emancipatory approach, I argued that discussing human enhancement through film can indeed provide a quality of its own. In bioethics, abstract talk is dominant, while movies about bioethics support a down-to-earth style of thinking and offer realistic but unreal images of possible future enhancement (see p. 184). In the section following, human enhancement at the movies was subsumed into doing science at the movies; cinematic traditions such as stock characters, stock settings, and stock props were pointed out, stressing that mad scientists have always been shown as obsessed with the goals of the transhumanist movement (see p. 184). The forms and effects of human enhancement on the screen formed the topic of the argumentation following this, and in the case of the norms of being human and the needs of human beings, a number of well known qualities were presented as essentially human. It is clear that history of philosophy has been used in film in order to define a human nature that is still illustrated through mirror foils of the technological age and improved through betterment of body and mind: whether through changing what we are made of or through uploading what we are (see p. 184). The societies presented in the movies were characterized predominantly as negative utopias, displaying the totalitarian characteristics of human social life that radicalize or symbolize social injustices, warning against a future society 'in which existing divisions will be exacerbated – in short, one that will be more unfair than our own' (Gavaghan 2009, p. 82). Showing undreamt-of possibilities of a depicted future science, the films at least call in question the status of today's average body functions and thus emphasize that distinctions like health/sickness and disability/non-disability are socially constructed and changeable. Simultaneously, they can be looked upon as utopian dreams about the possibilities of the individual. People are forced to surrender biological data to public observation, but have gained the talents that humankind has dreamt of for thousands of years. Long-desired traits such as the improvement of virtue, heart, and reason are afforded by genetic modification, so that the human properties that have traditionally been held responsible for a good life are developed to a high degree (see p. 184). Finally, it was argued that the movies investigated in this paper do not predict what will be feasible tomorrow as future science, but do initiate a cinematic debate about what is thinkable today as science fiction. Using various possibilities inherent in the medium, and inspired by both scientific knowledge and collective dreams and fears of technological progress, they offer a thought experiment. The sociological distinction between risk and danger was a helpful instrument to show that, according to these movies, no matter what ethical pros and cons are proposed, society will increasingly be forced to take serious decisions concerning human enhancement in general.

What is shown on the screen thus both indicates and co-constructs a debate about human enhancement. As soon as real science is aware of human nature as a problem, human enhancement as a possible solution is debated in public and reflected by art, which is then seen by a large audience. Obviously, as filmmakers take into account various ethical, artistic, and financial considerations – such as partisanship and influence, translation into cinematic forms, entertainment and profit – they hold a gatekeeping position, presenting certain arguments in certain ways and leaving other aspects completely unsaid. Although they are influenced by both groups' discourses, they in turn also influence both the lay audience's and the expert audience's problem awareness. To read the movies in a scholarly way thus means to realize how a story is developed out of the enhancement debate, to examine which ethical, economic, and psychological problems are chosen and which are concealed, and to relate all this to the purpose (if definable) of art.

## Notes

1. This is how a character describes the artificial beings in Ridley Scott's *Blade Runner* (0:21). All quotations from the movies discussed in this paper are referenced by the hour and minute in which they appear (h:mm).
2. Attempts to restore good health and/or good functioning have been common practice for hundreds of years, in the form of glasses, prostheses, or cardiac pacemakers. The boundaries between normal and abnormal extensions of the body are, however, increasingly blurred, so that even though the distinction between therapy and enhancement is pervasive, it is actually inappropriate.
3. Chan-wook Park's movie of the same name (released 2006) involves a girl who becomes convinced she that is a cyborg after experiencing a trauma.
4. For basic information consider, among others, Matias Martinez and Michael Scheffel. 2009. *Einführung in die Erzähltheorie*. 8th edn. Munich: Beck; David Bordwell and Kristin Thompson. 2008. *Film Art: An Introduction*. 8th edn. Boston: McGraw-Hill; Nicole Mahne. 2007. *Transmediale Erzähltheorie: eine Einführung*. Göttingen: Vandenhoeck & Ruprecht.
5. In his review of *Gattaca*, Gavaghan refers to Jude Law, Ethan Hawke, and Uma Thurman in the main roles. Scarlett Johansson and Ewan McGregor star in *The Island*, Christian Bale in *Equilibrium*, and Bruce Willis and Rosamund Pike play the protagonists in *Surrogates*. The *Matrix* trilogy was a remarkable exception when the movie was released, because Keanu Reeves' appearance references the male underground hacker scene, which is far from the image of prototypical Hollywood masculinity (cf. Matrix 2006, ch. 3).
6. The scientists' equipment at the beginning of *Avatar*, for example, resembles the tubes required for computerized tomography. Viewers may feel aggression or claustrophobia because they identify with the body caught in the machine.
7. The Prometheus myth is discussed by Trijsje Franssen in this volume.
8. Philip K. Dick's short stories and novels are a good example. Ridley Scott's *Blade Runner* (1982), which must be mentioned as a classic in terms of AI and its ethical concerns, is one of the movies based upon his texts. 'Replicants', who have stronger bodies and are 'at least equal in intelligence' (*Blade Runner* 0:02) compared to human beings, are used as working slaves in outer space. A group of them illegally return to earth, wanting to meet their human creator, a natural scientist, whom they force to reveal the meaning of

life. However, the boundaries between AI and human beings turn out to be increasingly unstable. Reading *Blade Runner* as portraying a mission to catch evil machines would be naïve, as the movie confronts the viewer with typical genre problems that are still recurrent today: artificial beings questioning their existence (cf. *The Island*), the ethical implications of consciousness (cf. e.g. *The Bicentennial Man*, 1999, directed by Chris Columbus), and debates about enhancing bodies. To what extent are human rights restricted to human beings? What is human if machines are, according to the advertising slogan broadcast in the movie, 'more human than human' (*Blade Runner* 0:21)? Are we allowed to switch off a machine that has developed self-consciousness (cf. in general Coleman and Hanley 2009)?

9. Speech in *Avatar* (0:09).
10. The opposite, though exceptional, is much more interesting as a work of art. An impressive example is the cinematic (and therefore mainly visual) adaptation of mathematical (and therefore mainly abstract) thinking in Darren Aronofsky's *Pi* (1998).
11. Speech in *Matrix Revolutions* (1:54).
12. But we should not forget that in former times, aging was also seen as improvement without enhancement: 'Growing [sic] old' was accompanied not solely by physical decline, but also by the attribution of wisdom and dignity, rights and honors. This tradition has disappeared.
13. Speech in *Gattaca* (0:45).
14. As a provocative act and as a remark on health selection in society, Berkeley University sent cotton buds to its applicants in 2010 – in real life (cf. Schmid 2010).
15. Speech in *The Island* (1:54).

## References

### Movies

Aronofsky, D. (1998): *Pi.*
Bay, M. (2005): *The Island.*
Cameron, J. (2009): *Avatar.*
Columbus, C. (1999): *The Bicentennial Man.*
Fincher, D. (2008): *The Curious Case of Benjamin Button.*
Forster, M. (2006): *Stranger Than Fiction.*
Kosinski, J. (2011): *TRON: Legacy.*
Lang, F. (1927): *Metropolis.*
Mostow, J. (2009): *Surrogates.*
Niccol, A. (1997): *Gattaca.*
Park, C. (2006): *I'm a Cyborg, But That's OK.*
Scott, R. (1982): *Blade Runner.*
Wachowski, A. and Wachowski, L. (1999): *The Matrix.*
Wachowski, A. and Wachowski, L. (2003): *The Matrix Reloaded.*
Wachowski, A. and Wachowski, L. (2003): *The Matrix Revolutions.*
Wimmer, K. (2002): *Equilibrium.*

## Literature

Bordwell, D. and Thompson, K. (2008): *Film Art: An Introduction*, 8th edn. Boston: McGraw-Hill.

Bühl, A. (ed.) (2009): *Auf dem Weg zur biomächtigen Gesellschaft? Chancen und Risiken der Gentechnik*. Wiesbaden: VS Verlag für Sozialwissenschaften.

Clayton, J. (2003): *Charles Dickens in Cyberspace: The Afterlife of the Nineteenth Century in Postmodern Culture*. Oxford: Oxford University Press.

Coleman, S. and Hanley, R. (2009): *Homo sapiens*, robots, and persons in *I, Robot* and *Bicentennial Man*. In: Shapshay, S. (ed.) *Bioethics at the Movies*. Baltimore, MD: Johns Hopkins University Press, pp. 44–55.

Dworkin, R. (2000): *Sovereign Virtue*. Cambridge: Harvard University Press.

Easton, L. (2008): Passing genes in *Gattaca*, or, straight genes for the queer guy. In: Easton, L. and Schroeder, R. (eds.) *The Influence of Imagination: Essays on Science Fiction and Fantasy as Agents of Social Change*. Jefferson, NC: McFarland, pp. 70–82.

Frayling, C. (2005): *Mad, Bad and Dangerous: The Scientist and the Cinema*. London: Reaktion.

Gavaghan, C. (2009): 'No gene for fate?' Luck, harm, and justice in *Gattaca*. In: Shapshay, S. (ed.) *Bioethics at the Movies*. Baltimore: Johns Hopkins University Press, pp. 75–86.

Grebowicz, M. (ed.) (2007): *SciFi in the Mind's Eye: Reading Science through Science Fiction*. Chicago: Open Court.

Hijikata, T. (ed.) (1997): *Riskante Strategien: Beiträge zur Soziologie des Risikos*. Opladen: Westdeutscher Verlag.

Kirby, D. A. (1999): The new eugenics in cinema: Genetic determinism and gene therapy in *Gattaca*. In: *Science Fiction Studies*, 26, pp. 193–215.

Luhmann, N. (1990): Risiko und Gefahr. In: *N. L.: Soziologische Aufklärung 5: Konstruktivistische Perspektiven*. Opladen: Westdeutscher Verlag, pp. 131–169.

Mahne, N. (2007): *Transmediale Erzähltheorie: eine Einführung*. Göttingen: Vandenhoeck & Ruprecht.

Maio, G. (2008): Das Motiv des Klonens im Fernsehen und Film. In: Schmidt, K. W., Maio, G. and Wulff, H. J. (eds.) *Schwierige Entscheidungen. Krankheit, Medizin und Ethik im Film*. Frankfurt/M.: Haag + Herchen, pp. 291–323.

Martinez, M. and Scheffel, M. (2009): *Einführung in die Erzähltheorie*, 8th edn. München: Beck.

Matrix, S. E. (2006): *Cyberpop: Digital Lifestyles and Commodity Culture*. New York: Routledge.

Michael, M. and Carter, S. (2001): The facts about fictions and vice versa: Public understanding of human genetics. In: *Science as Culture*, 10, pp. 5–32.

Münkler, H., Bohlender, M. and Meurer, S. (eds.) (2010): *Sicherheit und Risiko: über den Umgang mit Gefahr im 21. Jahrhundert*. Bielefeld: transcript.

Nordmann, A. (2010): Ethics beyond the power of imagination: A critique of speculations about future humans. *Lecture at the Conference Good Life Better – Anthropological, Sociological and Philosophical Dimensions of Enhancement*, Lübeck, 14 October 2010.

Perkowitz, S. (2007): *Hollywood Science: Movies, Science, and the End of the World*. New York: Columbia University Press.

Platzgummer, V. (2003): *Die Errettung der Menschheit. Studien zu den Science-fiction-Filmen 'Gattaca' und 'Matrix'*. Marburg: Tectum.
Putnam, H. (1981): *Reason, Truth, and History.* Cambridge: Cambridge University Press.
Schmid, T. (2010): Die Gier nach frischem Speichel. In: *taz*, 11 June 2010, http://www.taz.de/!53832/, last access 04.04.2014.
Seeßlen, G. (2000): Traumreplikanten des Kinos. Passage durch alte und neue Bewegungsbilder. In: Aurich, R. (ed.) *Künstliche Menschen, manische Maschinen, kontrollierte Körper.* Berlin: Jovis, pp. 13–45.
Shapshay, S. (ed.) (2009): *Bioethics at the Movies.* Baltimore: Johns Hopkins University Press.
Stacey, J. (2010): *The Cinematic Life of the Gene.* Durham: Duke University Press.
Weingart, P. (2005): Von Menschenzüchtern, Weltbeherrschern und skrupellosen Genies. Das Bild der Wissenschaft im Spielfilm. In: Weingart, P. (ed.) *Die Wissenschaft der Öffentlichkeit. Essays zum Verhältnis von Wissenschaft, Medien und Öffentlichkeit.* Weilerswist: Velbrück Wissenschaft, pp. 189–206.
Wulff, H. J. (2008): Gentechnik im Film. In: Schmidt, K. W. Maio, G. and Wulff, H. J. (eds.) *Schwierige Entscheidungen. Krankheit, Medizin und Ethik im Film.* Frankfurt/M.: Haag + Herchen, pp. 267–290.

# 11
# Transhumanism's Anthropological Assumptions: A Critique

*Nikolai Münch*

## Introduction

The question of whether humans can, may, or should be 'improved' through biotechnological means that go beyond the healing of diseases, and, if yes, how far and under which conditions, is a disputed one (e.g., Savulescu and Bostrom 2010; Coenen et al. 2010). This is unsurprising, insofar as the debate about enhancement in general seems to be about fundamental questions of what it means to be human. However, these anthropological arguments are particularly associated with the opponents of enhancement, such as Jürgen Habermas and Francis Fukuyama. In the following text, therefore, I want to show that the proponents of enhancement also engage with these fundamentally anthropological issues. I highlight a trend whose representatives have established themselves in recent years as the most vehement proponents of 'improving' humans: the transhumanists. Among the most influential representatives of transhumanism is the Swedish philosopher Nick Bostrom, who currently heads the Future of Humanity Institute at Oxford University. He is co-founder of the World Transhumanist Association (WTA), the umbrella organization of the transhumanists, and author of the *Transhumanist FAQ* (Bostrom 2003b), which provides a program for the WTA. The investigation below will concentrate on Bostrom's works; based on his role within the transhumanist movement, it is fair to assume that Bostrom can be taken as generally exemplary of the principles of transhumanism.

My objective is to show that it is not only the opponents of enhancement whose arguments are based on controversial anthropological claims but also those of Bostrom, as an example of one of enhancement's most radical supporters. It will involve elaborating on the concept of the human mind and body upon which Bostrom's argumentation is based, using his notion of 'uploading'. The result will be a demonstration of the fact that the transhumanist position, as represented by Bostrom, relies on an implausible conception of, and contempt for, the human body, and that this contempt serves as an indication of transhumanists' overarching view of the human.[1]

I will first describe the objectives, sources, and argumentation of Bostrom's transhumanism, then the focus will shift to the idea of uploading. It will become clear that this vision of enhancement is not built on Cartesian dualism, but on a computer functionalist description of the human. Then I will present in detail the consequences of computer functionalism for the view of the human body, which can be described in Nietzsche's words as 'despising the body'. To provide contrast, I will draw upon the body phenomenology of Maurice Merleau-Ponty, confronting Bostrom's uploading idea and his overarching argumentation with these phenomenological analyses.

## Transhumanism: Objectives, resources, and argumentation

The term 'transhumanism' was introduced in 1957 by the English biologist Julian Huxley, considered one of the pioneers of contemporary transhumanism (Huxley 1957; cf. Heil 2010, p. 129). Yet the term has only become established over the last 25 years, to describe the conquest of the limits of the human using rational scientific and technological means (Coenen 2009, p. 268). Although transhumanism is not based on a monolithic, closed set of ideas – even Nick Bostrom only speaks of a 'loosely defined movement' (Bostrom 2003a, p. 493)[2] – shared fundamental convictions can be recognized.

The starting point for transhumanist considerations is the conviction that humans, in their current (biological) state as *Homo sapiens*, are at a relatively early stage of their developmental transformation. Set against future possible developments, the current state of the human being is seen primarily as the *de facto* and contingent limitation of human possibilities and abilities (Bostrom 2003a, p. 494). When considering these limited human possibilities, which are thus open to improvement, transhumanists refer principally to four sectors. First, the lengthening of 'health-span' (Bostrom 2008, p. 108) – the time during which human beings are 'fit', both mentally and physically, unaffected by aging processes and diseases – should be extended as far as possible, preferably infinitely. Second, human cognitive ability should be increased as much as possible. This affects both general intellectual capabilities such as memory, concentration, and logical reasoning, and also more specific ones such as mathematical understanding (Bostrom 2008, p. 117; Bostrom and Roache 2011). A third objective is the increase of emotional capacities. Bostrom sometimes speaks of 'mood and personality enhancement' (Bostrom and Roache 2007, pp. 133–136). This term refers to the ability to enjoy life – feeling joy, fun, and sensual pleasure; expressing interest and enthusiasm; gaining equilibrium and empathy – while avoiding 'negative' emotions such as aggression, hate, contempt, and so on if they are not appropriate to the situation (Bostrom 2008, pp. 119–120). The fourth point is the increase in physical abilities. These abilities have some overlap with those that Bostrom subsumes under the area of health-span, such as the strengthening of the immune system and so on. We should also think in terms of abilities such as strength and endurance. Bostrom also considers human 'improvement' by means of new senses, such as sonar or magnetic orientation (Bostrom 2003c, p. 4).

In all these sectors, the transhumanists see humans as limited by the stage of their evolutionary development; this could be surmounted by the continuous improvement of the *conditio humana*. The means to transcend human limits are rational: applied science and technology. In particular, gene technology and nanotechnology, information technology, and artificial intelligence are seen as having great potential (Bostrom 2003b, pp. 7–20).

What differentiates transhumanism from other approaches in the enhancement debate is the (in principle) unlimited nature of such interventions – assuming the safety of the techniques employed. While more moderate proponents of technological human 'improvement' reject an unlimited increase in human abilities, because transcending human existence would be associated with alienation from deeply held values and experiences that are closely linked with this mode of existence (Agar 2010, pp. 12–13 and Chapter 9; cf. also Agar 2007) or with the danger of a qualitatively new social inequality (Gesang 2007, pp. 9, 50–51), transhumanists have no such qualms. Set against the limitations of enhancement relative to a human scale, however this is defined, is the explicit goal of technologically surmounting the limits of *Homo sapiens* through posthuman 'future beings whose basic capacities so radically exceed those of present humans as to be no longer unambiguously human by our current standards' (Bostrom 2003b, p. 6). The vision of uploading, probably the most extreme means of overcoming the current *conditio humana*, is therefore also a unique selling point of transhumanism in the enhancement debate. Before the uploading idea is further investigated in light of the body and physicality, let us first take a brief look at Bostrom's argumentation. How is the transhumanist vision of technologically transcending the human justified?

The possible benefits of employing the means Bostrom suggests are, from one point of view, 'obvious' (Bostrom 2003a, p. 498; Bostrom 2008, pp. 113, 116), as outlined in the four principal sectors above. Nevertheless, the question remains: are there no upper limits? Is achieving a posthuman state desirable? To generate a positive answer to these questions, Bostrom develops various argumentation strategies.

A first strategy consists of using descriptions to paint a picture of posthuman states and ways of living, and of the valuable paths along which a life with posthuman capabilities would proceed:

> Let us suppose that you were to develop into a being that has posthuman healthspan and posthuman cognitive and emotional capacities. At the early steps of this process, you enjoy your enhanced capacities. You cherish your improved health: you feel stronger, more energetic, and more balanced. [...] You also discover a greater clarity of mind. You can concentrate on difficult material more easily and it begins making sense to you. You start seeing connections that eluded you before. [...] You are able to sprinkle your conversation with witty remarks and poignant anecdotes. Your friends remark on how much more fun you are to be around. Your experiences seem more vivid. When you listen to music you perceive layers of structure and a kind of musical logic to which you were previously oblivious; this gives you great joy. You continue to find the gossip magazines you used to read amusing, albeit in a different way than before; but you discover that you can get more out of reading Proust and *Nature*. [...] You have just celebrated your 170th birthday and you feel stronger than ever. Each day is a joy. You have invented entirely new art forms, which exploit the new kinds of cognitive capacities and sensibilities you have developed. You still listen to music – music that is to Mozart what Mozart is to bad Muzak. You are communicating with your contemporaries using a language that has grown out of English over the past century and that has a vocabulary and expressive power that enables you to share and discuss thoughts and feelings that unaugmented humans could not even think or experience.
>
> (Bostrom 2008, pp. 111–112)

Bostrom asks: would such a life with increased capabilities not be 'better' than one without? At the same time, however, he skips over a number of controversial questions. For instance, is such a pleasure in every lived day really achievable through enhancement technologies? This descriptive strategy is not much more than an illustration of the *promise* that for Bostrom is associated with radical enhancement. It does, however, make clear the endpoint of Bostrom's reasoning: the radical enhancement he proposes is measured against the assumed positive consequences for well-being, and is thus legitimized. What counts is that there are no upper limits, that well-being could be ratcheted up further at will.

A second strategy aims less at the promises of radical enhancement than at the limited possibilities of human beings in their current constitution. Imagine, says Bostrom, the space of possible existence. This consists of combinations of capabilities or other general parameters of life, and – it is assumed – is limited only by the laws of physics. Only a small part of this physically defined space of possibility is accessible to animals – chimpanzees, for example – and much more to humans. So it is possible for humans to have valuable ways of living that are not available to chimpanzees. Thus it is plausible that the space that humans themselves cannot yet access also contains valuable ways of living – and these possible modes of living would be accessible (only) to posthuman beings (Bostrom 2008, p. 122; cf. also Bostrom 2003a, pp. 494–495 and Bostrom 2003c, pp. 2–3). Bostrom assumes that valuable ways of living can plausibly be described by – or founded upon – a set of capabilities or general parameters alone, with no other necessary factors to take into account.

The most important of Bostrom's argumentation strategies is, however, the third one. While the argument sketched above assumes that in the 'space of possibility' accessible only to posthuman beings more valuable ways of living await to be lived, the third line of reasoning links these ways of living with the ideals of current humanity. He wants to argue 'that some of *our* ideals may well be located outside the space of modes of being that are accessible to us with our current biological constitution' (Bostrom 2003a, p. 495). Bostrom wants to show that from our current evaluative standpoint there are good reasons for developing posthuman capabilities. It is not a matter of giving up our current values and ideals to justify surmounting humans in their existing constitution, but rather that the consistent pursuit of current values leads humans beyond themselves.

> Transhumanism does not require us to say that we should favor post-human beings over human beings, but that the right way of favoring human beings is by enabling us to realize our ideals better

> and that some of our ideals may well be located outside the space of modes of being that are accessible to us with our current biological constitution.
>
> (Ibid.)[3]

The reasons leading us into the posthuman area are the same as those that underlie practices that are already widely accepted; that is, those which are carried out 'in order to protect and expand life, health, cognition, emotional well-being, and other states or attributes that individuals may desire in order to improve their lives' (Bostrom and Roache 2007, p. 3). The protection and promotion of life, health, cognition, and emotional well-being (the four core areas of the enhancement that Bostrom proposes) are, he argues, values that can be read from individual behavior as much as from governmental and legal regulations. For example, both individually and societally, we provide enormous resources for education. Education, of course, aims to increase cognitive capacities, but the reach of these 'conventional' means is limited, since the realization of our values is only possible as far as our present biological constitution allows. With the new means that transhumanism hopes for, these limits could be surmounted. Thus it is the 'old values', pursued through new means, that will carry us beyond the boundaries of being human (cf. Bostrom 2008; Bostrom and Sandberg 2009; Bostrom and Roache 2011).

We could describe this type of reasoning as a 'continuity argument'. Bostrom uses such arguments not only to support his standpoint, but also to counter critics of enhancement; his aim is to circumvent limitations of therapy and enhancement which are based on criteria that opponents view as morally significant barriers (Bostrom and Roache 2007, pp. 1–3). The continuity argument fulfills the function of taking reasons out of the opponents' hands (Murray 2007, pp. 497–498).

I will return to these lines of argumentative reasoning. But first I turn to Bostrom's view of the human body and mind, as illustrated by the concept of uploading.

## Transhumanist dualism?

Uploading, also described as 'whole brain emulation' (Bostrom and Sandberg 2008), is a characteristic feature of transhumanism that is little discussed in the broader enhancement debate. 'Uploading [...] is the process of transferring an intellect from a biological brain to a computer' (Bostrom 2003b, p. 17).[4] This is about scanning the structure of the human brain and then transferring it to a computer, or replicating it there. According to Bostrom's idea, consciousness and identity would be completely maintained. Uploading would thus achieve a lengthening of the health-span and an increase of cognitive ability, since the upload would not be subject to the biological aging process, and could be restarted through regular updates as necessary, for example after an accident. 'Thus your lifespan would potentially be as long as the universe's' (Ibid., p. 18). In addition, says Bostrom, cognitive enhancements would be easier to carry out on computer hardware than on a biological brain: 'For instance, if you were running on a computer thousand times more powerful than the human brain, then you would think a thousand times faster' (Ibid.).

What does the uploading vision betray about the implicit view of the human being, body and mind, within transhumanist discourse? Critics taking a classical position in the body-soul debate connect the idea of uploading with Cartesian dualism:

> The model of transhumanism is, however, complete dematerialization. It carries the dualism of *res cogitans* and *res extensa* too far, in that it declares the latter to be formable at will, and wishes to 'liberate' the mind totally from the human body.
>
> (Coenen 2009, p. 274)[5]

The material dualist Descartes took the position that the world consisted of precisely two substances: *res extensa*, the physical substance, the essential characteristic of which is extension; and *res cogitans*, the immaterial spiritual substance, the essential attribute of which is thinking. For Descartes, the differentiation between these two substances is not conceptual or analytical; rather, the two are different from one another in reality. The mental thus becomes ontologically independent. Thoughts, feelings, or pains become a mode or state of the immaterial spiritual substance. The body, on the other hand, is thought of as an extended material object, which like all other such objects functions according to purely mechanical principles. Descartes' two-substance theory, in addition to providing a metaphysical underpinning for its mechanistic physics and physiology against the Aristotelian-scholastic tradition, also pursued the goal of securing the immortality of the soul. This appeared to be accomplishable only if the soul were not a property of the perishable body, but was ontologically independent of it (Descartes 1972, pp. 15ff., 61ff.; cf. Perler 2006, pp. 169–180; Beckermann 2008, pp. 4–7, 29–37). Although humans are beings composed of both substances, Descartes appears to be interested almost exclusively in *res cogitans*. Only the thinking substance appears directly accessible and recognizable, and only it is addressed as the 'self'; the human appears as a 'thinking thing' (Descartes 1972, p. 27).

This sketch of Cartesian dualism allows us to recognize the intuition that underlies the link made by critics of transhumanism between the concept of uploading and Cartesian mind-body dualism. It emerges from the fact that transhumanists want to separate the human mind, which they see as alone in defining consciousness and identity, from the perishable body, in order to overcome mortality and biological limitations. But despite the parallels, the transhumanist approach is unlike that of Descartes. The thesis of a transhumanist (substance) dualism is countered by their decidedly naturalistic fundamental position:

> Transhumanism is a naturalistic outlook. At the moment, there is no hard evidence for supernatural forces or irreducible spiritual phenomena, and transhumanists prefer to derive their understanding of the world from rational modes of inquiry, especially the scientific method.
>
> (Bostrom 2003b, p. 46)

The term *naturalism* has been used, historically and in the present, by many different philosophical approaches (Keil and Schnädelbach 2000, pp. 11–12). In connection with the philosophy of the mind, however, a naturalistic view is incompatible with Descartes' approach, notwithstanding what it is supposed to describe in detail. According to Descartes, in addition to physical things there must also be independent, non-physical things – *res cogitans* – and thus also irreducibly spiritual phenomena; precisely that which transhumanists reject. The concept of uploading, for example, is not premised on the presence of an immaterial spiritual substance. Rather, in Bostrom's case it is based on a computer functionalistic view of the human mind; something which has particular consequences for the conceptualization of the human body.

## The computer functionalistic view

Bostrom does not necessarily need an immaterial mental substance for the separation of the mind from the body in uploading. He refers instead to the argument of multirealizability (Bostrom and Sandberg 2008, p. 15; Bostrom 2003d, p. 244). This argument is one of the core theses of the functionalistic theory of the mind developed in the 1960s, notably by Jerry Fodor and Hilary Putnam. One main impetus for this development was the inadequacy of identity theory at the time. Put simply, identity theory claims that every type of mental state is identical with a type of physical state (see Beckermann 2008, pp. 98–141). Beings that have a different neurophysiology, such as animals, Martians, or computers, cannot therefore have the same mental state as humans. The main objection against identity theory is based on empirical research results that showed that particular mental states in different persons are not correlated with identical neuronal states, and that even in one and the same person, these correlations can change over the course of a life (Ibid., pp. 137–138). This suggests that particular mental states can be realized through different physical states (multirealizability). Functionalism takes up this argument.

The basic theses of functionalism can be briefly summarized as follows (Beckermann 2008, p. 142):

- Mental states are, by their nature, functional states.
- Functional states of a system are characterized solely by their causal role: through inputs from outside the system through which they are caused, by outputs that they cause outside the system, and by their causal relations to other states within the system.

The standard example for this type of functionally defined mental state is pain. Viewed functionally, pain is a mental state that fulfills a particular causal role; one can characterize it through its typical causes and effects. It is typically caused by an injury or something similar (input), it causes behaviors such as groaning or actions such as going to the doctor (output), and it causes the wish to remove the pain or to be distracted from it (causal relation to other system states). For functionalism, a being has the mental state of pain precisely when it is in a state that fulfills this causal role.

The functionalistic basis of uploading can also be read from Bostrom's description of its implementation (Bostrom and Yudkowsky 2011, pp. 10–11). In the first step, the parts of the brain functionally relevant to its operations are scanned. A second step adds the computational properties of these functionally relevant basic elements. This second step allows Bostrom's functionalism to be specified more precisely as computer functionalism. While functionalism in its general form, as described above, assumes that mental states are functional states (or put another way, that the mind is the functional organization of the brain), computationalism in its general formulation states that the functional organization of the brain consists of stepwise manipulation, directed by algorithms, of syntactically defined symbols with semantic properties. The two standpoints are logically independent of one another in these broad definitions. Computer functionalism combines both standpoints into the thesis that the mind is the computational organization of the brain (Piccinini 2010, pp. 270–271). Or put as a common (but rather vague) catchphrase: the computer functionalist views the mind as the software of the brain.

What, precisely, does computationalism add to functionalism? First, a computer can be understood as a system describable through its functional states alone. However, at a deeper level the computer becomes a paradigm. The computer is a symbol-processing machine, which creates and processes character strings using simple basic operations, the order of which is determined by an underlying algorithm. Bostrom's second uploading step aims at this computational level, since it involves adding 'the computational structure and the associated algorithmic behavior of its components' (Bostrom and Yudkowsky 2011, p. 10) to the previously scanned wetware circuit diagram. (Human) thinking appears, in this view, to be calculating, a rule-guided, formal manipulation of symbols, analogous to the way in which a Turing machine or a computer works, and can therefore in principle be replicated by suitable computers in all details and aspects: 'This is the basis for brain emulation: if brain activity is regarded as a function that is physically computed by brains, then it should be possible to compute it on a Turing machine' (Bostrom and Sandberg 2008, p. 7).

The idea of uploading is not based on Cartesian dualism, but is the result of a naturalization of the human mind through the computer functionalistic model. One could also say that the computer functionalistic view gives a mechanistic explanation of thinking (Dupuy 2009, pp. x–xi, 3–4), as thought processes can basically be duplicated mechanically using a Turing machine. On the premise that human thinking and consciousness apply structures that can be found everywhere in nature, the algorithms are materialized in one case by the Turing machine, and in the other by the biological human brain. In this sense the computer metaphor of the human mind continues the older machine metaphor, which was particularly popular in Enlightenment materialism. Furthermore, if one does not understand 'naturalism' in terms of the categories that produce the antithesis between nature and culture, 'then the computer metaphor of mind is in complete continuity with most contemporary efforts at naturalization' (Keil 1993, p. 149).

## Consequences of computer functionalism for the view of the body

A first consequence of Bostrom's version of computer functionalism could be described as brain-centerednes. This brain-centeredness is linked with what we understand as input and output within the functionalistic framework. As described above, mental states are defined by their causal roles, that is, events outside the system by which they are caused (inputs), outputs that they cause outside the system, and their causal relationships to other mental states inside the system. But what counts as an input or output? This question is important because the answer will also establish the system boundaries of the mind. Functionalism has, in principle, three possible answers (Beckermann 2008, p. 175). One possibility would be to define inputs as the electrochemical signals that the brain receives from the sensory organs. Analogously, outputs would be the electrochemical signals that the brain sends to the muscles. A second possibility would be to understand inputs as the physical stimuli that the sensory organs process (such as sound waves that reach the ear). Outputs would then be the movements of the limbs. A third version of establishing the system boundaries would be to extend the definition of the terms input and output to include the different environmental situations in which we find ourselves and the changes in the environment that our actions bring about. Bostrom argues for the first viewpoint: inputs and outputs are understood as electrochemical signals that the brain receives from or sends to the sensory organs. This view of input and output has been formulated explicitly: 'The emulation produces and receives neural signals corresponding to motor actions and sensory information' (Bostrom and Sandberg 2008, p. 30).

If the system boundary of the mind is drawn so tightly, the actual logical subject of mental states is the brain. This would mean that it is not Franz who feels pain, but his brain,[6] for this is where the subject is situated. Moreover, it should make no difference to mental states, including their phenomenal quality,[7] exactly how the electrochemical signals arrive and what they cause. Whether the brain is attached to a mechanical (robot) body, whether it is connected to a camera instead of eyes, whether it even has a 'worldly' body at all or rather the signals are created through the simulation of a virtual world, is irrelevant to the mental state (Bostrom and Yudkowsky 2011, p. 11), since all this is external to the mental system. Causal influence is exercised only by the neuronal or electrochemical signals that the brain receives (as a physical realization). Equally, whether the output signals control a biological body, a robot body, or 'just' a virtual body, makes no difference, since the only output that matters in the causal system of the mind is the signal that leaves the brain.

If the body is outsourced and upstream of the subject in this way, it becomes, together with all the sensory organs, a tool to provide the groundwork for the conscious 'self', which is exclusively realized in the brain. It becomes clear that Bostrom understands the body and sensory organs primarily as tools, as contingent appliances for processing input and output; for example, when he describes physical enhancement analogously to the use of 'external tools' (Bostrom and Roache 2007, p. 11). If the body is purely an external tool for the subject, then in principle there is almost no difference between the use of a forklift truck or of the tool 'body'. If the body (whether mechanical or biological) is not an integral component of the subject, it is not particularly worthy of protection either. An intervention into the human body does not, therefore, require any special justification.

A second consequence that is closely linked with 'brain-centeredness' is a kind of 'antibiologism'. Beyond the determination of input and output, there is an additional aspect of functionalism that particularly concerns the human biological body, and especially the brain. '[T]he computer model of the mind has a built-in antibiological bias' (Block 1990, p. 391), which is associated with the implication of multirealizability, described above. Since what is decisive about mental states is their functional role, the physical substrate through which the functional organization operates is irrelevant. From this viewpoint it is therefore obvious that the physical basis of the mind can be judged according to how well it fulfills 'its function', that is, the realization of the mind. Bostrom leaves little doubt that the 'biological hardware' of a human – that is, the brain – comes off badly in this evaluation: 'The three-pound, cheese-like thinking machine that we lug around in our skulls can do some neat tricks, but it also has significant shortcomings' (Bostrom 2003b, p. 3).

From a transhumanist perspective then, the body appears to be a purely contingent tool of input gathering or output processing. The body is linked with the 'self' only causally, as a component of the causal chain that provides input information and processes output information. In addition, it is not an integral component of the conscious subject, which is adequately described simply by the mental states realized in the brain. The body can, to an extent, be neglected and (re)formed at will, without fundamentally changing any elements of human experience or human identity. Such a reforming or turning off of the body might even be desirable, for although the biological human body fulfills its causally specified functions, it only just does so. It suffers from a substantial flaw: its unreliability and perishability.

This kind of view resembles that of the 'despisers of the body', against whom Nietzsche's Zarathustra preaches (Nietzsche 1988, pp. 39–41). To those who despise the body as 'simple material', who see only its instrumental worth and who value the purely spiritual, he counters: 'Body am I entirely, and nothing more' (Ibid., p. 39). Zarathustra insists that the body is not simply the opposite of the spirit, but rather designates the whole being of the human: every self-experience is physical, every experience of the world is mediated physically. 'The body thus proves itself to be the fundamental *a priori*, behind which we can in no way go back without confirming it again (*Der Leib erweist sich somit als das grundlegende Apriori, hinter das auf keine Weise zurückgegangen werden kann, ohne daß es zugleich wieder bestätigt wird*)' (Pieper 1990, p. 150).[8]

## Despising the body or a phenomenology of corporeality? Merleau-Ponty as a contrast

As a contrast to such 'despising of the body', Merleau-Ponty's phenomenology will now be drawn upon, and in particular his magnum opus *The Phenomenology of Perception* (Merleau-Ponty 2002). Merleau-Ponty is not alone in calling to mind the human body as a meaningful field of philosophical reflection; others have also referred to the irreducibly physical dimension of human existence, as the example from Nietzsche shows. Yet Merleau-Ponty counts as 'something like the patron saint of the body' (Shusterman 2005, p. 151). Indeed, 'none can match the bulk of rigorous, systematic, and persistent argument that Merleau-Ponty provides to prove the body's primacy in human experience and meaning' (Ibid.). In what follows, I show that the human body cannot be neglected, but is a basal dimension of human existence – a dimension that characterizes our relationship to ourselves just as fundamentally as our relationship to the world.

Merleau-Ponty's starting point is the phenomenon of perception. Engaging with the philosophical tradition, he demonstrates the irreducible corporeality of perception. He chooses a phenomenological approach: not an objectified view from outside, but using an understanding of the phenomenon of perception from the first person perspective to make it understandable how perception reveals a world. Perception is therefore considered primarily as a fundamental aspect of being-in-the-world (*être-au-monde*). Corresponding to perception, this being-in-the-world is fundamentally corporeal. Below, I will concentrate on two connected points of Merleau-Ponty's corporeal phenomenology and attempt to show how they prove Bostrom's computer functionalistic view to be implausible. Concretely, one can show through Merleau-Ponty's reasoning that the world, as it comes into being for us through perception, is fundamentally based on corporeal rules (Waldenfels 1986, p. 159); and also, that the body functions as a 'natural self' and is thus an integral component of the concrete self (Waldenfels 1986, p. 161). These two points, which concern the human relationship to the world and to oneself, undermine some of the consequences of Bostrom's computer functionalistic view, namely:

(1) the instrumental view of the human body;
(2) the assumption of brain-centeredness; and
(3) equating thinking with calculating.

1. First, the purely instrumental view of the human body, which Bostrom's version of computer functionalism entails, is implausible against a backdrop of Merleau-Ponty's phenomenology. Treating the body as a pure object of the world, which for contingent reasons is closer to the human mind because it is located further 'behind' in the causal chain of input reception, does not do justice to the role of the body. With Merleau-Ponty's phenomenology we can show that the body does not serve merely as a tool of input reception, locomotion, and so on. Rather, our being-in-the-world is irreducibly corporeal.

How can we show that the body is constitutive of our being-in-the-world? It is essential to Merleau-Ponty that perception is always meaningful: 'Vision is already inhabited by a meaning (sens) which gives it a function in the spectacle of the world and in our existence' (Merleau-Ponty 2002, p. 60). According to him, perceiving something means 'grasping' something. The registration of meaningless sensory data, in the sense of empiricism, does not do justice to this (Ibid., pp. 3ff.), for ultimately what we find in our experience is neither atomistic qualitative morsels of the outside world, abstracted from our perceived coherent environment, nor completed objects (cf. Carman 2008, pp. 45–53; Good 1998, pp. 36–42). Nor can intellectualism adequately describe our perception experience (Merleau-Ponty 2002, pp. 30ff.), for explicitly propositional judgments rest on a more basic form of understanding that does not require the application of simply any concept (cf. Carman 2008, pp. 53–61; Good 1998, pp. 42–50). What precedes and forms the basis of the possibility of qualities isolated from perception or the formulation of explicit judgments, according to Merleau-Ponty, is the 'phenomenal field' (Merleau-Ponty 2002, p. 62). 'This phenomenal field is no "inner world", the "phenomenon" is not a "state of consciousness", or a "mental fact"' (Ibid., p. 66); rather, it is 'that aspect of the world always already carved out and made available and familiar to us by our involuntary bodily perceptual capacities and unthinking behaviors' (Carman 2008, p. 64).

We are thus always in a meaningful world, not in the sense that the water is in the glass or the dress is in the wardrobe (cf. Heidegger 2006, p. 54), but in a more existential way: the subject cannot, in contrast to the water in the glass, be described and understood exhaustively at all, independent of this field of meaning. 'That is, if we want to say how it is with a person, give his "state-description", our characterization has to take in some features of the world which surrounds him. Not just any features; we have to take in those features which have meaning for him' (Taylor 1989, p. 1). It is this field of meanings that makes the subject what it is.

This meaning-laden phenomenal field, which encounters us in our perception, is constituted by our body. Merleau-Ponty thus speaks of the body as the 'vehicle of being in the world' (Merleau-Ponty 2002, p. 94). The phenomenal field appears as a meaningful field, which has always been structured by the structures of our body and its own possibilities and capabilities (Carman 2008, p. 106). It is constituted by the sensorimotor structures, capabilities, and possibilities of the body. 'The structure of perception, we might say, just *is* the structure of the body' (Carman 2008, p. 81, emphasis in original).

Nevertheless, we may not understand this corporeal perspectivity as a subjective deformation of perception (Merleau-Ponty 1965, p. 186); on the contrary, this corporeally anchored perspectivity of the phenomenal field is the condition for our having a meaningful perception field at all, without losing orientation. The drawback here is the openness to the world. The body 'limits our comprehension while making it possible in the first place' (Bermes 2004, p. 74). Merleau-Ponty thus designates the phenomenal field as being also a 'transcendental field' (Merleau-Ponty 2002, p. 71). This indicates that it concerns not only the causal effects that the body or the position and properties of our sensory organs have on our perception. To say that perception is irreducibly corporeal does not mean that the constitution of my eyes makes me unable to see around corners, but that the kind of perception we have is essentially corporeal: the nature, and not just the concrete contents of our experience, is determined by our corporeal constitution. Therefore, descriptions of the situations in which we are only make sense against a backdrop of corporeality; they give the conditions of intelligibility (Taylor 1995a, pp. 62–63; cf. also Taylor 1995b). Thus perception – and associated with it, the human relationship and connectivity to the world – cannot be understood if the concrete corporeal situations in which humans find themselves are ignored. But this is precisely what Bostrom's computer functionalistic view requires.

A phenomenological analysis of perception shows, therefore, that we always perceive the world to be meaningful. Without this meaning-laden field, the subject cannot adequately be described. But the meaningful world, as it appears in our perception, is indissolubly linked to the human body, because it is the condition for the possibility that we perceive the world as meaningfully as we do. Bostrom's computer functionalistic view cannot capture this level of human corporeality, which far exceeds the purely instrumental function of the body, and thus also misses the corporeally transmitted human understanding of the world and oneself.

2. Closely associated with this is the brain-centeredness of computer functionalism. For Bostrom, consciousness, the personal self, is completely identical to the functions of the brain. Uploading does not produce anything different. All aspects of the human mind, including personal identity, should be maintained. In Merleau-Ponty's view, however, the human spirit and consciousness are not just that which the brain does. The subject or 'self' cannot be exclusively located in the brain, because the structure of the body already provides a framework for the structure of our meaningful perception. The theoretical framework underlying the idea of uploading simply abstracts from the body and thus also from the specific corporeal embedding of the human in the world. Without this corporeal being-in-the-world, however, the human phenomenal experience – whatever physiological or physical bases it has – cannot be described or understood. But it is precisely these phenomenal experiences that need to be retained throughout the process of uploading because, as Bostrom's argument has shown, he is mainly concerned with subjectively experienced well-being.

This point is underlined through what Merleau-Ponty terms the 'natural self'. The body not only structures our relationship to the world, but is also constitutive of human self-understanding. In some way, self- and world-relationships are two sides of the same coin, 'for if it is true that I am conscious of my body *via* the world [...], it is true for the same reason that my body is the pivot of the world' (Merleau-Ponty 2002, p. 94). But what does this mean for our self-understanding? If our being-in-the-world is essentially corporeal, and a description of the subject cannot be abstracted from its embedding in the world, does this not mean that the body is an integral component of the 'self'? Merleau-Ponty draws precisely this conclusion. He speaks of the body as the 'natural self', which is 'as it were, the subject of perception' (Merleau-Ponty 2002, p. 239). What does this mean?

Our behavior contains a layer of the instinctive and unconscious, which is not based on purely physiological reflexes, nor is it completely transparent and available to us. For example, a color does not appear as an unambiguous physical given, but as a chromatic form in a visual field, with particular expressive values, and laden with cultural meanings. I discover all of this, but this discovering is not something entirely free, which I can decide to do and from which I can always refrain. This discovering is in some way self-generating. We could speak of 'improper intentionality' (Waldenfels 1986, p. 160) at work here. There is already sense and significance in the perception of color, yet it also requires my active composition. Underlying the personal subject, which is transparent to itself, is a layer of corporeal experience, which is already laden with meaning. 'So, if I wanted render precisely the perceptual experience, I ought to say that *one* perceives in me' (Merleau-Ponty 2002, p. 250, emphasis in original). The relative anonymity of the original perception experience, which is Merleau-Ponty's target here, equates with this level of the 'natural self': 'Conscious behavior is supported, carried, stimulated everywhere by corporeal impulses, which provide a sense and in which the Self already dwells, instead of just using them instrumentally' (Waldenfels 1986, p. 161). The body at a pre-personal level has thus always been integrated into the personal 'self'.

Bostrom provides a conception of the human being that is abstracted from its corporeal constitution. He cannot therefore put this level of the 'natural self' into perspective. Irrespective of whether, for example, an upload would have consciousness at all, if one follows Merleau-Ponty it would not have a world of human experiences and meanings because the computer functionalistic basis underlying uploading is unable to capture this phenomenal level even theoretically. Insofar as uploading is based on computer functionalism, as argued by Bostrom, the promise of a replication of consciousness in all its phenomenal facets is implausible, because the theoretical concept simply lacks the means to capture and explain this phenomenal experience.[9]

3. How can one use Merleau-Ponty to show that thinking is not identical to calculating? The computer functionalistic model grasps thinking as a rule-guided, formal manipulation of symbols. This implies a representational model. The basic idea is that the substance of the subjective experience of the representational content is a state that is directed towards a particular part of the world (Metzinger 2006, p. 315). This part of the world is again depicted or represented internally for the subject of consciousness. This explains how mental states are or could be intentional states; put simply, that they relate to *something*, are directed towards an object or a content (one thinks something, one loves *something*, and so on) (cf. Beckermann 2008, pp. 291–300). The computer manipulates chains of symbols (character strings of bits) according to a particular algorithm. These strings, for their part, are the representation of the objects that are to be calculated, meaning the numbers, which the computer cannot manipulate as abstract entities. According to representationalism, this is also what occurs in the human mind.

This picture is based on the idea that a state or a property 'in' our mind is responsible for creating a link to 'external' objects. In principle this 'inner' property of the subject can be isolated from its situative and corporeal embedding. However, if one takes Merleau-Ponty's phenomenological descriptions seriously, it becomes obvious that there is already a preconceptional 'improper intentionality', which is closely linked to human corporeality, and which cannot be reduced to a representational picture (Carman 2008, pp. 35–36; more detail in Taylor 2005).

## Conclusion

In the transhumanist picture, computer functionalism is the foundation for the idea of uploading. The extension of human cognitive performance is also modeled by computer functionalistic information processing (see Bostrom and Roache 2011). Here, too, some skepticism derived from the thinking of Merleau-Ponty is appropriate.

More important than these issues of concrete enhancement measures, however, is the overall approach to conceptualizing humans within the transhumanist picture. In Bostrom's argumentation described above, the human appears as a bundle of physically describable capabilities. This objectivized perspective of the third person is what provides the framework within which an unproblematic continuity, from chimpanzees via *Homo sapiens* up to transhuman beings, can be postulated. Only the physical laws count, but these remain the same for all. This continuity underlies Bostrom's arguments. What this perspective loses sight of is that within this physical continuum – at least in Bostrom's computer functionalistic approach – there may well be discontinuities and qualitative leaps that affect the level of phenomenal experience, the perspective of the first person.

The phenomenology of Merleau-Ponty can be used to detect such possible fractures. For example, in perception a meaningful phenomenal field appears to a human, a field which is also irreducibly formed by the constitution of that person's body. But this field is transcendental in the sense described above: it forms the framework through which the world becomes meaningfully accessible to us. What world would reveal itself in perception to beings with new sensory capabilities such as sonar (Bostrom 2003c, p. 4) or beings with eyes in the back of their heads? It would not be comparable with the human world: it would not only be incomparable in the quantitative sense, because we lack the extent of the sensory information that these beings would have; this world would be a fundamentally different one, because with the radical changing of the human body the transcendental framework would also change, and thus the fundamental level to which the meaningfully accessed human world relates its significance. Therefore, there is no ongoing continuum ad infinitum, but a qualitative fracture. And if so, how can one then make consistent statements about the possibly higher well-being of transhuman beings?

## Notes

1. Roduit, Baumann and Heilinger (2013) show that in the enhancement debate 'bioconservatives' as well as 'bioliberals' rely on specific ideas of human perfection. These different ideas of human perfection seem to be based on corresponding conceptions of human nature. In the case of Bostrom his contempt of the bodily aspects of the human condition informs his vision of uploading as one step towards human perfection.
2. An overview of the themes and discussions of the transhumanist movement is given by the texts in More and Vita-More (2013).

3. This is also the reason why Bostrom considers his transhumanistic philosophy to be conservative at its core:

   > I therefore do not regard my claim as in any strong sense revisionary. On the contrary, I believe that the denial of my claim would be strongly revisionary in that it would force us to reject many commonly accepted ethical beliefs and approved behaviors. I see my position as a conservative extension of traditional ethics and values to accommodate the possibility of human enhancement through technological means.
   >
   > (Bostrom 2008, p. 113)

4. In addition to the 130-page long roadmap for 'whole brain emulation' (Bostrom and Sandberg 2008), Bostrom's uploading idea is also considered in Bostrom 2003c, p. 4, Bostrom 2005, pp. 7–12, and Bostrom and Yudkowksy 2011, pp. 10–11.
5. A similar direction is indicated in Krüger (2004, p. 198) and Hopkins (2012, p. 232). Krüger's position is, however, ambivalent. Although he recognizes that the transhumanists he investigates have functionalistic motives, he assigns this to their presumed dualism. He also emphasizes this by drawing transhumanism's line of tradition back to Descartes (Ibid., pp. 178–180). As we shall see, however, linking functionalism and Cartesian dualism is implausible.
6. This should not, however, be misunderstood biologically, as the brain is, according to the functionalist view, a more or less contingent realization of the actual functional organization (multirealizability).
7. The term 'phenomenal quality' here refers to how it is like for you to undergo or have a certain experience.
8. If we view this aspect of corporeality in Zarathustra, it appears inappropriate to compare the transhumanist movement with Nietzsche and his Superman (Sorgner 2009). Hauskeller (2010) is also critical of this thesis. Bostrom too sees only superficial commonalities between transhumanism and Nietzsche's thinking (Bostrom 2005, pp. 4–5).
9. A further question is whether giving up a brain-centered model does justice to phenomenal experience. This is attempted, sometimes with reference to Merleau-Ponty, by approaches such as inactivism or the extended mind thesis (for an overview, see e.g. Thompson and Stapleton 2009). The latter at least is also enlisted to legitimize enhancement measures (e.g., Levy 2007). Whether these approaches can do justice to phenomenal experience, as described by Merleau-Ponty, is controversial. Carman (2008, pp. 225–229) and Dreyfus (2007) are critical of this.

## References

Agar, N. (2007): Whereto transhumanism? The literature reaches a critical mass. In: *Hastings Center Report*, 37(3), pp. 12–17.

Agar, N. (2010): *Humanity's End. Why We Should Reject Radical Enhancement.* Cambridge, MA: The MIT Press.

Beckermann, A. (2008): *Analytische Einführung in die Philosophie des Geistes*, 3rd edn. Berlin: de Gruyter.

Bermes, C. (2004): *Maurice Merleau-Ponty zur Einführung*, 2nd edn. Hamburg: Junius.
Block, N. (1990): The mind as the software of the brain. In: Smith, D. E. and Osherson, D. N. (eds.) *An Invitation to Cognitive Science. Volume 3: Thinking*, 2nd edn. Cambridge, MA: The MIT Press, pp. 377–425.
Bostrom, N. (2003a): Human genetic enhancements: A transhumanist perspective. In: *The Journal of Value Inquiry*, 37, pp. 493–506.
Bostrom, N. (2003b): *The Transhumanist FAQ. A General Introduction*, version 2.1, World Transhumanist Association, http://www.transhumanism.org/resources/FAQv21.pdf.
Bostrom, N. (2003c): *Transhumanist Values*, http://www.nickbostrom.com/ethics/values.pdf.
Bostrom, N. (2003d): Are we living in a computer simulation? In: *The Philosophical Quarterly*, 53(211), pp. 243–255.
Bostrom, N. (2005): *History of Transhumanist Thought*, http://www.nickbostrom.com/papers/history.pdf.
Bostrom, N. (2008): Why I want to be a posthuman when I grow up. In: Gordijn, B. and Chadwick, R. (eds.) *Medical Enhancement and Posthumanity*. Berlin: Springer, pp. 107–137.
Bostrom, N. and Roache, R. (2007): Ethical issues in human enhancement. In Ryberg, J., Petersen, T., Wolf, Clark (eds.) *New Waves in Applied Ethics*. Basingstoke: Palgrave Macmillan, pp. 120–152.
Bostrom, N. and Roache, R. (2011): Smart policy: Cognitive enhancement and the public interest. In: Savulescu, J., Meulen, R. and Kahane, G. (eds.) *Enhancing Human Capacities*. Malden: Wiley-Blackwell, pp. 138–149.
Bostrom, N. and Sandberg, A. (2008): *Whole Brain Emulation: A Roadmap*, www.fhi.ox.ac.uk/reports/2008-3.pdf.
Bostrom, N. and Sandberg, A. (2009): Cognitive enhancement: Methods, ethics, regulatory challenges. In: *Science and Engineering Ethics*, 15(3), pp. 311–341.
Bostrom, N. and Yudkowsky, E. (2011): The ethics of artifical intelligence. In: Ramsey, W. and Frankish, K. (eds.) *Cambridge Handbook of Artificial Intelligence*. Cambridge: Cambridge University Press.
Carman, T. (2008): *Merleau-Ponty*. London, New York: Routledge.
Coenen, C. (2009): Transhumanism. In: Bohlken, E. and Thies, C. (eds.) *Handbuch Anthropologie. Der Mensch zwischen Natur, Kultur und Technik*. Stuttgart: Metzler, pp. 268–276.
Coenen, C., Gammel, S., Heil, R. and Woyke, A. (eds.) (2010): *Die Debatte über 'Human Enhancement'. Historische, philosophische und ethische Aspekte der technologischen Verbesserung des Menschen*. Bielefeld: transcript.
Descartes, R. (1972): *Meditationen über die Grundlagen der Philosophie mit den sämtlichen Einwänden and Erwiderungen*. Hamburg: Meiner.
Dreyfus, H. L. (2007): Why Heideggerian AI failed and how fixing it would require making it more Heideggerian. In: *Philosophical Psychology*, 20(2), pp. 247–268.
Dupuy, J.-P. (2009): *On the Origins of Cognitive Science. The Mechanization of the Mind*. Cambridge, MA: The MIT Press.
Extropy Institute (2003): *Transhumanist FAQ*, http://www.extropy.org/faq.htm, date accessed 14 April 2011.
Gesang, B. (2007): *Perfektionierung des Menschen*. Berlin: de Gruyter.

Good, P. (1998): *Merleau-Ponty. Eine Einführung*. Düsseldorf, Bonn: Parerga.

Hauskeller, M. (2010): Nietzsche, the overhuman and the posthuman: A reply to Stefan Sorgner. In: *Journal of Evolution and Technology*, 21(1), pp. 5–8.

Heidegger, M. (2006): *Sein und Zeit*, 19th edn. Tübingen: Max Niemeyer.

Heil, R. (2010): Trans- und Posthumanism. Eine Begriffsbestimmung. In: Hilt, A., Jordan, I. and Frewer, A. (eds.) *Endlichkeit, Medizin und Unsterblichkeit. Geschichte – Theory – Ethik*. Stuttgart: Steiner, Ars moriendi nova 1, pp. 127–149.

Hopkins, P. D. (2012): Why uploading will not work, or, the ghosts haunting transhumanism. In: *International Journal of Machine Consciousness*, 1(4), pp. 229–243.

Huxley, J. (1957): Transhumanism. In: *New Bottles for New Wine*. London: Chatto & Windus, pp. 13–17.

Keil, G. (1993): *Kritik des Naturalismus*. Berlin, New York: de Gruyter.

Keil, G. and Schnädelbach, H. (2000): Naturalismus. In: *Naturalismus. Philosophische Beiträge*. Frankfurt/M.: Suhrkamp, pp. 7–45.

Levy, N. (2007): Rethinking neuroethics in the light of the extended mind thesis. In: *The American Journal of Bioethics*, 7(9), pp. 3–11.

Merleau-Ponty, M. (1965): *The Structure of Behavior*. London: Methuen.

Merleau-Ponty, M. (2002): *Phenomenology of Perception*. London, New York: Routledge.

Metzinger, T. (2006): Repräsentionalistische Theorien des Bewusstseins I. Einleitung. In: *Grundkurs Philosophie des Geistes. Vol. 1: Phänomenales Bewusstsein*. Paderborn: Mentis, pp. 315–316.

More, M. and Vita-More, N. (eds.) (2013): *The Transhumanist Reader: Classical and Contemporary Essays on the Science, Technology, and Philosophy of the Human Future*. Chichester: Wiley-Blackwell.

Murray, T. H. (2007): Enhancement. In: Steinbock, B. (ed.) *The Oxford Handbook of Bioethics*. Oxford: Oxford University Press, pp. 491–515.

Nietzsche, F. (1988): *Also sprach Zarathustra*, 2nd edn. Berlin: de Gruyter.

Perler, D. (2006): *René Descartes*, 2nd edn. Munich: Beck.

Piccinini, G. (2010): The mind as neural software? Understanding functionalism, computationalism, and computational functionalism. In: *Philosophy and Phenomenological Research*, 81(2), pp. 269–311.

Pieper, A. (1990): *'Ein Seil geknüpft zwischen Tier und Übermensch'. Philosophische Erläuterungen zu Nietzsche's erstem 'Zarathustra'*. Stuttgart: Klett-Cotta.

Roduit, J. A. R., Baumann, H. and Heilinger, J.-C. (2013): Human enhancement and perfection. In: *Journal of Medical Ethics*, 39(10), pp. 647–650.

Savulescu, J. and Bostrom, N. (eds.) (2010): *Human enhancement*. Oxford, New York: Oxford University Press.

Shusterman, R. (2005): The silent, limping body of philosophy. In: Carman, T. and Hansen, M. B. (eds.) *The Cambridge Companion to Merleau-Ponty*. Cambridge: Cambridge University Press, pp. 151–180.

Sorgner, S. (2009): Nietzsche, the overhuman, and transhumanism. In: *Journal of Evolution and Technology*, 20(1), pp. 29–42.

Taylor, C. (1989): Embodied agency. In: Pietersma, H. (ed.) *Merleau-Ponty. Critical Essays*. Boston, London: University Press of America, pp. 1–21.

Taylor, C. (1995a): Lichtung or Lebensform. Parallels between Heidegger and Wittgenstein. In: *Philosophical Arguments*. Cambridge, MA; London: Harvard University Press, pp. 61–78.

Taylor, C. (1995b): The validity of transcendental arguments. In: *Philosophical Arguments*. Cambridge, MA; London: Harvard University Press, pp. 20–33.

Taylor, C. (2005): Merleau-Ponty and the epistemological picture. In: Carman, T. and Hansen, M. B. (eds.) *The Cambridge Companion to Merleau-Ponty*. Cambridge: Cambridge University Press, pp. 26–49.

Thompson, E. and Stapleton, M. (2009): Making sense of sense-making: Reflections on enactive and extended mind theories. In: *Topoi*, 28, pp. 23–30.

Waldenfels, B. (1986): Das Problem der Leiblichkeit bei Merleau-Ponty. In: Petzhold, H. (ed.) *Leiblichkeit. Philosophische, gesellschaftliche und therapeutische Perspektiven*. Paderborn: Junfermann, pp. 149–172.

# 12
# Be Afraid of the Unmodified Body! The Social Construction of Risk in Enhancement Utopianism

*Sascha Dickel*

The current debate about enhancement is shaped by implicit hopes and fears for the future. To better understand the ongoing debate it might be helpful to turn our attention to visions that make these hopes and fears explicit. Hence, in recent years a new field of inquiry has emerged that focuses on the analysis of human enhancement futures (see, e.g., Coenen et al. 2010; Dickel 2011; Gordijn and Chadwick 2009). This chapter is part of these efforts to improve our understanding of futuristic visions of enhanced humans, through reconstructing the function of risk in these futures of human enhancement.

Visions of enhancement typically come in two forms: utopias and dystopias. Dystopian futures paint a posthuman future in gloomy colors. They have been made popular by science fiction movies such as *Gattaca* (Niccol 1997),[1] but can also be found in academic and political writings (Habermas 2003; Kass 2004; McKibben 2003). Dystopians try to convince us that taking the road to enhancement is very dangerous: they claim, for example, that genetically enhancing ourselves or our children might lead to grave social injustices, if humanity were to be divided into an enhanced class of 'GenRich' and an unenhanced class of 'GenPoor' (Silver 1998). Furthermore, they affirm that altering our minds and emotions through the use of drugs may threaten our dignity (Fukuyama 2002), and that modifying our bodies with cybernetic implants might compromise our humanity (Agar 2010).

Critics of enhancement frame the imminent biotechnological transformation of humankind not as an inevitable fate, but as a risk that might be prevented if we make the right decisions in the present. This emphasis on risk is not very surprising, because the pragmatic meaning of dystopias is to warn their audience, and warning only makes sense if a danger can be averted.

Much more surprising is the fact that it is not only critics but also proponents of enhancement (e.g., transhumanists) who use explicit or implicit warnings in their writings. Both dystopias and utopias (which promote enhancement) use the construction of risk as an important discursive strategy, but utopias turn the critics' notion of risk upside down: enhancement utopias frame the decision *not to enhance* as more risky than enhancement itself.

Before I analyze this discursive strategy, a deeper look at utopianism and risk in the context of social theory is needed. After that I will turn to enhancement utopianism itself and present six different types of risk construction in this kind of utopian thinking. The chapter concludes with some general remarks on (enhancement) utopianism and society.

## Future imperfect

When we think about utopias we often think of works of fiction. But we are also used to talking about the utopian aspirations of social movements, utopian manifestos, or even utopian pictures. All these are cases of communication. They could be read as messages of meaning. For the purpose of this chapter I will ignore all the differences of genres and media. Instead I will focus on the message of utopias. I claim that all utopias are communicated visions of a desired future that is in some way radically different from the present (for a deeper theoretical discussion see Dickel 2011).

Modern utopian thinking emerged with a new social understanding of time. According to the historian Reinhart Koselleck (1989), the future was closed during the Middle Ages. People believed that the future of humanity was determined by God and that the world would end soon. The time of apocalypse was also regarded as the end of secular time itself. But with the arrival of the modern era, the concept of the future shifted. People became aware of the fact that things were changing rather drastically – and that these changes were happening because of human actions. With the development of modern society, religion lost its power over the future, and a new perception of futurity began. Modernity started to believe in an open future (Koselleck 1989).

An open future means that radically different paths to the future seem possible. Since early modernity, human action has been viewed as an important force that could change the course of history. An open future implies contingency and contingency leads to uncertainty. The emergence of an open future makes it increasingly necessary for individuals and societies alike to decide their way into the future: if many paths are possible, which road should be taken?

Utopias provide an answer to this question. They essentially tell us that there may well be different paths to the future, but only one way will lead to a radically better future. Typical utopias paint images of perfection. We imagine this state of perfection to be reached when core values such as equality, liberty, or wealth have been fully realized. But no one has ever arrived in utopia. Utopia is like a rainbow, which sometimes seems to be just around the next corner but ultimately is impossible to reach. Utopias are thus always *present* futures (Luhmann 1976). Their function is to serve as a guide for action in the present – not to be realized (Ricoeur 1986).

But in our time, which sociologists such as Zygmunt Bauman and Anthony Giddens term 'late modernity' (Giddens 1991), utopia no longer seems to be a trusted guide for action. We have watched utopian hopes come and go, and we now perceive utopias of the Renaissance and the Enlightenment as part of our history, no longer as pictures of the future; we witnessed the fall of socialist states, and chuckle about dreams of the coming Age of Aquarius. The sociological interpretation of utopias as present futures has reached the collective consciousness of late modernity, and this deconstruction has caused utopias to lose much of their power (Levitas 2007; Nassehi 1994). Instead, the dominant way of thinking about the future today is in terms of risk (Nassehi 1994; Reith 2004).

Social theorists of our time provide an illuminating interpretation of the emergence of our utopia-skeptical age. They describe late modernity as an age of accelerating change (Koselleck 2003; Lübbe 1998; Rosa 2003, 2005) and claim that the current level of social acceleration undermines utopian thinking. Social acceleration can be defined 'by an increase in the decay-rates of the reliability of experiences and expectations' (Rosa 2010). Acceleration therefore increases the contingency of the future. Its effect is that we can no longer rely on our past experiences, and we must expect that our hopes for the future will be disappointed.

Acceleration is a key feature of modernity in general and was an important precondition for the emergence of an open future, because the perception of change undermined static conceptions of the future. But acceleration has now reached a new level:

> Early modernity promised the capacity to shape and control world and time and to initiate and historically legitimate future progress.

> But in late modernity, time itself has come to destroy the potential for any form of social or substantial control, influence, or steering.
>
> (Nassehi 1993, p. 375, cited and translated by Rosa 2003, p. 22)

Because of the vast number of simultaneous events around the globe that influence each other in unforeseeable ways, an increased rate of innovation, and a perception of ever-changing values and trends, the hope of controlling the course of history is largely gone. The future of late modernity is still open. However, society seems unable to structure its own future – not because of a belief in predestination but because of the experience of a radical contingent future beyond control. This also means a shift in the identity of the actors. The dynamics of late modernity call for reflexive and flexible subjects who expect their identity to change over time. Typical subjects of late modernity do not expect to be able to predict what they will desire in ten years' time, and it is even more difficult for them to decide on the best of all possible worlds for the next generation, because they cannot predict how their values, goals, and dreams might change (Bauman 2001; Rosa 2005). Hence, the new social consensus in our uncertain time seems to be: do not try to achieve the best of all possible worlds but prevent the worst – and the idea of preventing the worst (for oneself and society as a whole) increasingly relies on the concept of risk:

> Although its meanings have changed since its emergence in the seventeenth century, and although it is used in a variety of ways to describe different social situations today, the concept of risk can still be defined largely through its attempt to calculate and so manage the uncertainties of the future. It is an expression of the likelihood of some situation or event – usually negative – occurring, and so, when we are talking about risks, we are talking about the future.
>
> (Reith 2004, p. 386)

In the world of late modernity, the ongoing calculation of risk seems to be the most viable strategy for moving into the future. Acting in the awareness of risk has become synonymous with rationality itself. It may be impossible to know what will happen, but if things do go wrong, we still have the comfort of knowing that we acted rationally – that is, in a risk-rational manner (Nassehi 1994; Reith 2004). By acting in risk-rational way we try to prepare ourselves for an unknown future.

But utopia and risk are not necessarily mutually exclusive constructions of futurity. The main theoretical thesis of this chapter is that in late modernity the incorporation of risk constructions into utopian images of the future provides a powerful discursive strategy for new kinds of utopian thinking. That is, utopias that do not strive for perfection but for increased flexibility, better realization of opportunity, and improved survival chances in the face of accelerating change.

## The risks of non-enhancement

Enhancement utopianism is an illuminating case. Like all other utopians, enhancement utopians express the desire for a radically different future – for themselves, for other individuals, and for society as a whole. They believe that this utopian future will be realized through technologies such as drugs, genetic engineering, and cybernetic implants, which might be used not only for therapeutic purposes but also to improve the healthy human body in some way. Although some of these technologies have already arrived, more radical enhancement technologies (such as genetic manipulation that could double human intelligence, or brain chips that might allow mind reading) are largely speculative (Coenen et al. 2010; Dickel 2011; Gordijn and Chadwick 2009). The application of these radical enhancement technologies should, according to enhancement utopians, transform the human condition into something better – and as the examples below will demonstrate, this 'better' has nothing to do with the old utopian endeavor for perfection.

Many utopian texts that try to convince us of the wonders of posthuman futures have their origin in transhumanist writings. In transhumanism, enhancement utopias take the form of a biopolitical movement strongly tied to futurist thinking. Transhumanists believe that technology will soon enable us to transform our minds and bodies in order to radically increase our intelligence and eliminate aging. Many transhumanist organizations, such as the global transhumanist umbrella association Humanity Plus and several institutes (inside and outside universities) led by transhumanists, have emerged over the last two decades. Some transhumanists, such as the inventor Ray Kurzweil or the biogerontologist Aubrey de Grey, have even become well-known media figures.

We also find enhancement utopias in the bioethical discourse of liberal eugenics. Defenders of liberal eugenics argue that parents should have the right to manipulate the genes of their children through genetic engineering, or should at least be able to select the best possible child through preimplantation genetic diagnosis (PGD). Not every defender of liberal eugenics is a utopian. But in the writings of the main supporters of liberal eugenics, such as Julian Savulescu, John Harris, and Jonathan Glover, we always encounter the utopian image that biotechnology will allow us to overcome our biological limitations and direct our own evolution.

Transhumanism is more interested in self-transformation and cybernetic enhancement, while liberal eugenics is focused on the genetic enhancement of generations to come. However, the two discourses are closely linked and there is an increasing overlap, both institutionally and thematically.

Much has been written about the philosophical standpoints of liberal eugenics (Buchanan et al. 2000; Kollek 2005) and transhumanism (Bostrom 2001; Kettner 2006; Siep 2006). Their social and historical context has also been reconstructed quite extensively (Bostrom 2005a; Coenen et al. 2010; Dickel 2011). I will not repeat these findings here. Instead I will focus on how enhancement utopianism constructs non-enhancement as a risky decision, using six examples.

## The risk of the genetic lottery

My first example is from the film *Gattaca*. Although this example is fictional we can learn much about the discourse strategies of enhancement utopianism from it. In *Gattaca*'s dystopic vision of a genetically enhanced society, we encounter a couple (Maria and Antonio) who visit a geneticist who performs *in vitro* fertilization; he confronts them with their 'genetic responsibility'. The doctor thus takes on the role of a genetic utopian (for real-life examples see Savulescu 2007; Stock 2003).

GENETICIST: You've already specified blue eyes, dark hair and fair skin. I have taken the liberty of eradicating any potentially prejudicial conditions – premature baldness, myopia, alcoholism and addictive susceptibility, propensity for violence and obesity.

MARIA (interrupting, anxious): We didn't want – diseases, yes.

ANTONIO (more diplomatic): We were wondering if we should leave some things to chance.

GENETICIST (reassuring): You want to give your child the best possible start. Believe me, we have enough imperfection built-in already. Your child doesn't need any additional burdens. And keep in mind,

> this child is still you, simply the *best* of you. You could conceive naturally a thousand times and never get such a result.
>
> (Niccol 1997)

In the above discussion with the geneticist, the parents must evaluate their future child against the backdrop of the *best possible* child. This changes their relationship to their future child fundamentally. They are now accountable for every decision they take. Even if they decide to 'leave some things to chance', they have to regard this as a choice, and therefore as a risk that they have voluntarily taken (cf. Kollek 2005). Proponents of liberal eugenics contrast the possibility of choosing the best possible child via PGD with the perils of natural conception. They argue that conceiving naturally is like daring to take part in a 'genetic lottery' (Glover 1984, p. 47), instead of taking responsibility for your child's genetic constitution.

Choosing your children's traits through PGD is still more or less a utopian dream (or, depending on who you ask, a dystopian nightmare). Nevertheless, some proponents of liberal eugenics and transhumanism actually speculate about further biomedical interventions, most notably the possibility of not only selecting but also modifying your children's genes (or your own). In doing so, they argue, it may become possible to leave the realm of mere human possibilities behind.

## The risk of restriction to human opportunities

Some authors in favor of enhancement reason that we should only use enhancement technologies (genetic enhancement or enhancement by other means) to become or create 'better humans' (Miller and Wilsdon 2006a), and not to transform humans into something else (Agar 2010). However, transhumanist philosophers such as Nick Bostrom argue otherwise:

> The range of thoughts, feelings, experiences, and activities accessible to human organisms presumably constitute only a tiny part of what is possible. There is no reason to think that the human mode of being is any more free of limitations imposed by our biological nature than are those of other animals. In much the same way as Chimpanzees lack the cognitive wherewithal to understand what it is like to be human – the ambitions we humans have, our philosophies, the complexities of human society, or the subtleties of our relationships with one another, so we humans may lack the capacity to form

> a realistic intuitive understanding of what it would be like to be a radically enhanced human (a 'posthuman') and of the thoughts, concerns, aspirations, and social relations that such humans may have. Our own current mode of being, therefore, spans but a minute subspace of what is possible or permitted by the physical constraints of the universe [...]. It is not farfetched to suppose that there are parts of this larger space that represent extremely valuable ways of living, relating, feeling, and thinking.
>
> (Bostrom 2003, p. 1)

Bostrom takes several important argumentative steps in constructing the idea of 'staying human' as a risk. First, he presents several alternative forms of being; second, he claims that these posthuman forms of beings are something we could actually become; third, he suggests that a posthuman form of life would be desirable because these enhanced beings could be smarter, stronger, and have a far longer life-span. By presenting the notion of becoming posthuman as a realistic option, Bostrom transforms staying human into a decision that individuals could take – at the risk of missing the posthuman opportunities.

Genetic engineering and cybernetic implants could, as transhumanists argue, some day provide us with superhuman strength and intelligence, and even give us new senses or access to new performance abilities (e.g., playing the piano with 12 fingers, like a musician in a scene in *Gattaca*).

But even if we follow Bostrom's line of reasoning, we could still doubt the rationality of developing technologies that some future generations might use in order to become posthuman. Why should we care, if we (and our children) will die before having the chance to explore this posthuman realm? Maybe there is a way to solve this problem: radical life extension.

## The risk of death

A significant amount of enhancement utopianism deals with the issue of life extension (for an overview see Knell 2009). The most interesting thing about life extension in the context of this chapter is how it reframes aging and even dying as risks. Aging and dying are usually regarded as fates that we all must face sooner or later. Even the arrival of anti-aging products has not changed this fundamental expectation. Most current anti-aging treatments are only cosmetic, and even the more extreme promises of life extension techniques (e.g., calorie restriction) promise only to slow down aging a bit.

But biogerontologists such as Aubrey de Grey and organizations such as the Immortality Institute go one step further. Their 'mission is to conquer the blight of involuntary death' altogether (Immortality Institute 2007; see also Bostrom 2005b). They claim that science may soon uncover the mechanisms of aging and make unlimited lifespans possible – if the international research community would only pay more attention to biogerontological research (de Grey 2004; Miller and Wilsdon 2006b). Focusing on other research topics – or spending money on things other than aging research – is therefore framed as a risk, for it leads, proponents argue, to the unnecessary deaths of millions of people (for a critical assessment of arguments such as this, see Nordmann 2007).

But according to authors such as Ray Kurzweil and Terry Grossman, every individual could at least try to 'live long enough to live forever' (Kurzweil and Grossman 2004), even though the envisioned breakthroughs in biogerontology will not happen in the near future. They expect that in the latter half of the 21st century, nanotechnology will give humans the power to reshape their bodies as they wish and 'cure' death – so it might be enough to hang on for the next few decades, without being killed by an accident, crime, or fatal disease, in order to have the chance to become practically immortal.

And if all else fails, there is still something you can do, there is still a very easy option to decide against death – making a contract with the Alcor Life Extension Foundation in Arizona. Alcor uses the technology of cryonics to give people the (highly disputed) chance to cheat death.

> Cryonics is the speculative practice of using cold to preserve the life of a person who can no longer be supported by ordinary medicine. The goal is to carry the person forward through time, for however many decades or centuries might be necessary, until the preservation process can be reversed, and the person restored to full health.
>
> (Alcor Life Extension Foundation 2010)

Currently Alcor has nearly 1,000 clients and over 100 'patients' – speaking clearly, the corpses of dead people in cryonic tanks. As a client of Alcor, you invest in a kind of life insurance that should guarantee that you will end up in a cryonic tank after you have died, in the hope that you will be brought back to life by advanced nanotechnology at some point in the future.

Alcor is not the only cryonics organization around. There are (a few) other cryonics institutes that offer a cryonic preservation service, and there are also several other organizations that take care of transporting clients who have recently died (see Cryonics Institute 2010).

Most people in the scientific community consider cryonics a prime example of fringe science, as even transhumanists such as Bostrom are ready to admit (Bostrom 2005a, p. 10). But several transhumanists have nevertheless signed a contract or have said that they will do so in the near future. They consider this a rational decision, because, according to Bostrom, 'even a 5% or 10% chance of success could make an Alcor contract a rational option for people who can afford it and who place a great value on their continued personal existence' (Bostrom 2001).

As a matter of fact one of the most important intellectual fathers of nanotechnology, Eric Drexler, regarded cryonics as a 'door to the future' that would allow us to experience the coming age of nanotechnological wonders (Drexler 1986) – one of the many connections between the futuristic discourses of imagined enhancement technologies and visions of nanotechnology.

## The risk of overwhelming complexity

The preceding three forms of discourse strategies operate more or less on the level of the individual. But the example of life extension activists who call for the investment of more money into aging research has already shown that individual risks could easily be linked to social issues and collective decision-making. The following three strategies go one step further. They focus primarily on the social consequences of the decision not to enhance, whatever the individual risks might be.

In many of the writings of enhancement utopians you will find the argument that human beings are somehow unable to cope with the complexity of the cultural world they have created. It is suggested that our mammalian brains are not ready for the demands and challenges of the present, and as our world becomes more and more complex the gap between our biological and our socio-technical evolution will increase even further. Certainly, we can improve our cognitive capacities through learning, but sooner or later, it is argued, we will reach a plateau that represents the maximum realization of human potential. However, this will not stop our socio-technical world from growing ever more complex. We could therefore find ourselves increasingly unable to manage the risks that our cultural and technological environment pose.

You might have guessed the proposed solution in enhancement utopias: the transformation of our biological framework in order to manage the challenges of the 21st century (and centuries to come).

Proponents of enhancement admit that this might also be risky, but as the enhancement utopian and cryonics evangelist Robert Ettinger phrased it: 'Our culture is changing so fast that in order to cope with it, men must soon change also. To go forward is to risk disaster, but to stand still is to ensure it' (Ettinger 1989 [1972], p. 6).

## The risk of amoral acts

While the complexity argument focuses on the cognitive limitations of the human mind, a related argument highlights the moral flaws of human nature. In his book *What Sort of People Should There Be?* (1984), which is considered one of the founding works of liberal eugenics, British philosopher Jonathan Glover writes:

> In optimistic moments, we may hope that our history of cruelty and killing is part of a primitive past, to be left behind as civilization develops. But the events of our own century do not suggest that this process has gone far. [...] We are familiar with the effects of napalm, and not surprised by the daily use of torture in many parts of the world. Less dramatically, but with similar terrible effects, we are used to our own passivity in the face of so much hunger, disease and poverty.
>
> (Glover 1984, p. 181)

Glover claims that there seems to be a tendency in humans to use the gifts of technology against one another, just like the ape-men in the film *2001: A Space Odyssey* (Kubrick 1968) used their first tool – a bone – to slay other members of their tribe. In Glover's book, the explanation for this tendency is biological:

> To the extent that we are the product of our genes, we are to be explained [...] as survival machines for genes. [...] There is no reason why the qualities we value should exactly coincide with those which have led to gene survival. Perhaps, in terms of our values, the world would be a better place if people were more altruistic and generous than a perfectly calculated survival strategy for genes would make them.
>
> (Glover 1984, p. 181)

It goes without saying that Glover suggests that this biological problem can only be fixed with biotechnological means – namely genetic enhancement – because, in his reading of history, other utopian aspirations to create a kind of new human through institutional reforms or political revolutions have failed.

With arguments like this, the individualistic discourse of liberal eugenics, which usually focuses on the relationship between parent and child, turns into a highly political discourse. If human nature is constructed as the cause of human atrocities, and enhancement technologies are viewed as the only way to change human nature, relinquishing enhancement is turned into a risky decision, on the individual as well as the collective level: if we decide to stay away from enhancement technologies, we could be held responsible for all future atrocities that may result from an inherent amorality in human nature (see also Persson and Savulescu 2008).

## The risk of extinction

My final example comes again from the writings of Bostrom. In his 2007 paper *The Future of Humanity*, Bostrom discusses four scenarios for the long-term future of our species. The first scenario he calls 'recurrent collapse'. This means that humanity will go through a 'cycle of indefinitely repeating collapse and regeneration' (Bostrom 2007, p. 14). Bostrom claims it is very unlikely that this cycle will go on forever, because this would presuppose unusually stable environmental and cultural conditions. The second possibility is that the biological, cultural, and technological evolution of humanity will some day reach a plateau: the end of history. Bostrom argues that evolution and history show us that such stagnation is also not very probable. He considers it much more likely that we will join the fate of 'an estimated 99.9 % of all species that ever existed on Earth' (Bostrom 2007, p. 9): extinction. We could slowly die out, blow ourselves to pieces, or be killed by some form of cosmic catastrophe. According to Bostrom, the extinction of humankind is in the long run a very probable scenario.

But he also envisions an alternative to this rather gloomy prospect: posthumanity. In this scenario, humans would transcend their biology by means of enhancement, colonize space, and improve their condition indefinitely. In Bostrom's view, the 21st century constitutes a turning point in human history, whereby revolutionary technological changes could lead either to extinction or posthumanity. If we manage to reach posthumanity in the coming decades, Bostrom argues, we might prevent the risk of extinction and enter a phase of ongoing self-transformation.

## Enhancement utopianism as an expression of late modernity

Despite all their differences, the six examples discussed above have something in common: their discourse could be described as a 'riskization' of human nature. The term 'riskization'– originally developed by Olaf Corry in the context of security studies (Corry 2010) – means that an object that is normally regarded as unchangeable is constructed as being in some way malleable and subject to action. Additionally, the object itself is constructed as harmful or intolerable.

In utopias of enhancement, the natural way of bringing forth children, being human, the fact of dying, the cognitive limits of our nature, all the crime and cruelty in the world, and the prospect that some day humanity will become extinct, are presented as mere possibilities. Enhancement utopianism thus transforms things that we would usually regard as natural and inevitable into contingent objects; the use of enhancement technologies is framed as an achievable alternative. Even though enhancement utopias refer to long-term horizons, they always call on us to act quickly in order to make this alternative possible and prevent the risks of non-enhancement.

In light of the enhanced human of the future, present humans might appear somehow 'disabled',[2] although the opposite of, or counterpoint to, this 'disability' is not an ideal of 'perfect' functioning. In the first part of this chapter I suggested that typical utopias paint images of perfection. Enhancement utopianism does not, however, aim for perfection, but for never-ending improvement. In the world of these utopias, humans (and even posthumans) are always in need of improvement. Assessed from the imagined futures of enhancement utopianism, the state of 'disability' is an anthropological constant that no enhancement can fix, because – according to the logic of enhancement utopianism – some kind of improvement is always better than sticking to the status quo: if you want to live forever, there can never be enough protection against fatal threats. If you do not want to miss posthuman opportunities, you had better be prepared to improve your body and brain indefinitely. And because you never know where your new capacities could lead you and how your identity and your desires may be shaped over time, the logic of enhancement utopianism suggests acting in a way that unlocks as many options as possible.

The claims and aims of enhancement utopianism might seem provocative or even outrageous to some, but nevertheless we should take them seriously – as symptoms of our current age. I argue that enhancement utopianism is simply an exaggerated expression of a general mode of thinking in late modernity. It is not farfetched to claim that its rationality of unlocking options in the face of known and unknown risks is also a central rationality of late modernity itself (Nassehi 1999; Rosa 2005; Schulze 2000).

The sociologist and utopian scholar Ruth Levitas has argued that:

> [U]topia is the expression of desire for a better way of living. Whatever we think of particular utopias, we learn a lot about the experience of living under any set of conditions by reflecting upon the desires which those conditions generate and yet leave unfulfilled. For that is the space which utopia occupies.
>
> (Levitas 1990, p. 8)

So what can we learn from enhancement utopianism? In the theoretical section of this chapter I described the temporal orientation of current society in terms of acceleration, innovation, and risk. In such a society, enough is never enough. Late modernity is a society that calls for fluid identities, the willingness to always work on yourself and to improve your capacities in a way that will allow you to handle all foreseeable and unforeseeable kinds of risks. Such a society may no longer have any place for the old utopian ideas of perfection, because achieving a state of perfection would imply that changes are no longer necessary. A society that favors never-ending improvement also finds it difficult to acknowledge absolute limits such as death or extinction – conditions where further improvement seems impossible.

Enhancement utopianism reflects these key features of late modernity by pushing them to extremes and projecting them into the open and contingent future. It can therefore be interpreted as a cultural answer to a social dilemma: our society itself generates the expectation that never-ending improvement is possible and desirable, yet people have the experience that social and natural conditions set limits on improvement projects, and are confronted with the fact that our life-span itself is limited. In opposition to this contradiction of late modernity, utopias of human enhancement project a future where never-ending improvement is indeed possible. With their constructions of risk, they frame the present state of limitations as a mere option and confront us with the possibility of deciding against the limited present for the limitless future. Enhancement utopias thus fill the gap between the expectations and experiences of late modernity.

This cultural function makes them powerfully and techno-scientifically seductive.

Unlike social utopias that construct alternatives to the current form of society, enhancement utopias target our bodies and minds, which appear to be malleable and improvable objects. At the same time, these utopias do not question the basic foundations of modern society. Instead, they aspire to be a guide for our late modern age. Hence, we should not assume that utopias of enhancement will vanish soon – even if the assumed breakthroughs in enhancement technologies may never happen.

## Notes

1. See also Chapter 10 in this volume for a discussion of *Gattaca* and other science fiction films that deal with fictional transhumanist and posthumanist future scenarios.
2. See also Chapter 9 in this volume for further discussion of how the contemporary non-enhanced human may come to be seen as 'disabled' in comparison to enhanced posthumans.

## References

### Movies

Kubrick, S. (1968): *A Space Odyssey*.

Niccol, A. (1997): *Gattaca – Screenplay*, http://ethics.sandiego.edu/courses/Phil321/Handouts/Gattaca%20script–highlighted.pdf, date accessed 29 August 2011.

### Literature

Agar, N. (2010): *Humanity's End: Why We Should Reject Radical Enhancement*. Cambridge, MA: The MIT Press.

Alcor Life Extension Foundation (2010): *About Cryonics*, http://www.alcor.org/AboutCryonics/index.html, date accessed 29 August 2011.

Bauman, Z. (2001): *Liquid Modernity*. Cambridge: Polity Press.

Bostrom, N. (2001): *What is Transhumanism?* http://www.nickbostrom.com/old/transhumanism.html, date accessed 23 September 2010.

Bostrom, N. (2003): Transhumanist values. In: *Review of Contemporary Philosophy*, 4, pp. 87–101.

Bostrom, N. (2005a): A history of transhumanist thought. In: *Journal of Evolution and Technology*, 14.

Bostrom, N. (2005b): The fable of the dragon tyrant. In: *Journal of Medical Ethics*, 3, pp. 273–277.

Bostrom, N. (2007): *The Future of Humanity*, http://www.nickbostrom.com/papers/future.pdf, date accessed 26 August 2011.

Buchanan, A. E., Brock, D. W., Daniels, N. and Wikler, D. (2000): *From Chance to Choice: Genetics and Justice.* Cambridge: Cambridge University Press.

Coenen, C., Gammel, S., Heil, R. and Woyke, A. (eds.) (2010): *Die Debatte über 'Human Enhancement': Historische, philosophische und ethische Aspekte der technologischen Verbesserung des Menschen.* Bielefeld: transcript.

Corry, O. (2010): *Securitization and 'Riskization': Two Grammars of Security*, working paper prepared for Standing Group on International Relations, 7th Pan-European International Relations Conference, Stockholm, http://stockholm.sgir.eu/uploads/Risk%20society%20and%20securitization%20theory%20SGIR%20paper.pdf, date accessed 29 August 2011.

Cryonics Institute (2010): *Comparing Procedures and Policies*, http://www.cryonics.org/comparisons.html, date accessed 29 August 2011.

de Grey, A. (ed.) (2004): Strategies for engineered negligible senescence: Why genuine control of aging may be forseeable. In: *Annals of the New York Academy of Sciences*, 1019.

Dickel, S. (2011): *Enhancement-Utopien: Soziologische Analysen zur Konstruktion des Neuen Menschen.* Baden-Baden: Nomos.

Drexler, E. (1986): *Engines of Creation: The Coming Era of Nanotechnology.* New York: Anchor Books.

Ettinger, R. (1989): *Man into Superman*, http://www.cryonics.org/book2.html, date accessed 29 August 2011.

Fukuyama, F. (2002): *Our Posthuman Future: Consequences of the Biotechnology Revolution.* New York: Picador.

Giddens, A. (1991): *Modernity and Self-Identity. Self and Society in the Late Modern Age.* Stanford, CA: Stanford University Press.

Glover, J. (1948): *What Sort of People Should there Be? Genetic Engineering, Brain Control and their Impact on Our Future World.* Harmondsworth, Middlesex: Penguin.

Gordijn, B. and Chadwick, R. (eds.) (2009): *Medical Enhancement and Posthumanity.* Heidelberg: Springer.

Habermas, J. (2003): *The Future of Human Nature.* Cambridge: Polity.

Immortality Institute (2007): *Immortality Institute Constitution & Bylaws*, http://www.imminst.org/mission, date accessed 29 August 2011.

Kass, L. (2004): *Life, Liberty and the Defense of Dignity: The Challenge for Bioethics.* San Francisco: Encounter Books.

Kettner, M. (2006): Transhumanismus und Körperfeindlichkeit. In: Ach, J. S. and Pollmann, A. (eds.) *No Body Is Perfect: Baumassnahmen am menschlichen Körper. Bioethische und ästhetische Aufrisse.* Bielefeld: transcript, pp. 111–130.

Knell, S. (2009): *Länger Leben? Philosophische und biowissenschaftliche Perspektiven.* Frankfurt/M.: Suhrkamp.

Kollek, R. (2005): From chance to choice? Selbstverhältnis und Verantwortung im Kontext biomedizinischer Körpertechniken. In: Bora, A., Decker, M., Grunwald, A. and Renn, O. (eds.) *Technik in einer fragilen Welt: Die Rolle der Technikfolgenabschätzung.* Berlin: Ed. Sigma, pp. 79–90.

Koselleck, R. (1989): *Vergangene Zukunft: Zur Semantik geschichtlicher Zeiten.* Frankfurt/M.: Suhrkamp.

Koselleck, R. (2003): *Zeitschichten: Studien zur Historik*. Frankfurt/M.: Suhrkamp.

Kurzweil, R. and Grossman, T. (2004): *Fantastic Voyage: Live Long Enough to Live Forever*. Emmaus: Rodale Press.

Levitas, R. (1990): *The Concept of Utopia*. New York: Allan.

Levitas, R. (2007): For Utopia: The (limits of the) Utopian function in late capitalist society. In: Goodwin, B. (ed.) *The Philosophy of Utopia*. London: Routledge, pp. 25–43.

Lübbe, H. (1998): Gegenwartsschrumpfung. In: Backhaus, K. (ed.) *Die Beschleunigungsfalle oder der Triumph der Schildkröte*. Stuttgart: Schäffer-Poeschel, pp. 129–164.

Luhmann, N. (1976): The future cannot begin: Temporal structures in modern society. In: *Social Research*, 43, pp. 130–152.

McKibben, B. (2003): *Enough: Staying Human in an Engineered Age*. New York: Times Books.

Miller, P. and Wilsdon, J. (eds.) (2006a): *Better Humans? The Politics of Human Enhancement and Life Extension*. London: Demos.

Miller, P. and Wilsdon, J. (2006b): The man who wants to live forever. In: Miller, P. and Wilsdon, J. (eds.) *Better Humans? The Politics of Human Enhancement and Life Extension*. London: Demos, pp. 51–58.

Nassehi, A. (1993): *Die Zeit der Gesellschaft: Auf dem Weg zu einer soziologischen Theorie der Zeit*. Opladen: Westdeutscher Verlag.

Nassehi, A. (1994): No time for Utopia: The absence of Utopian contents in modern concepts of time. In: *Time & Society*, 3, pp. 47–78.

Nassehi, A. (1999): Das Problem der Optionssteigerung: Überlegungen zur Risikokultur der Moderne. In: Nassehi, A. (ed.) *Differenzierungsfolgen: Beiträge zur Soziologie der Moderne*. Opladen: Westdeutscher Verlag, pp. 29–48.

Nordmann, A. (2007): If and then: A critique of speculative nano ethics. In: *NanoEthics*, 1, pp. 31–46.

Persson, I. and Savulescu, J. (2008): The perils of cognitive enhancement and the urgent imperative to enhance the moral character of humanity. In: *Journal of Applied Philosophy*, 25, pp. 162–177.

Reith, G. (2004): Uncertain times: The notion of 'risk' and the development of modernity. In: *Time & Society*, 13, pp. 383–402.

Ricoeur, P. (1986): *Lectures on Ideology and Utopia*. New York: Columbia University Press.

Rosa, H. (2003): Social acceleration: Ethical and political consequences of a desynchronized high-speed society. In: *Constellations*, 10, pp. 3–33.

Rosa, H. (2005): *Beschleunigung: Die Veränderung der Zeitstrukturen in der Moderne*. Frankfurt/M.: Suhrkamp.

Rosa, H. (2010): Full speed burnout? From the pleasures of the motorcycle to the bleakness of the treadmill. The dual face of social acceleration. In: *International Journal of Motorcycle Studies*, 6(1).

Savulescu, J. (2007): Genetic interventions and the ethics of enhancement of human beings. In: Steinbock, B. (ed.) *The Oxford Handbook of Bioethics*. Oxford: Oxford University Press, pp. 516–535.

Schulze, G. (2000): *Die Erlebnis-Gesellschaft: Kultursoziologie der Gegenwart*. Frankfurt/M.: Campus.

Siep, L. (2006): Die biotechnische Neuerfindung des Menschen. In: Ach, J. S. and Pollmann, A. (eds.) *No Body Is Perfect: Baumassnahmen am menschlichen Körper. Bioethische und ästhetische Aufrisse.* Bielefeld: transcript, pp. 21–42.

Silver, L. M. (1998): *Remaking Eden: How Genetic Engineering and Cloning Will Transform the American Family.* New York: Avon Books.

Stock, G. (2003): *Redesigning Humans: Choosing Our Genes, Changing Our Future.* Boston: Houghton Mifflin.

# Index

Note: The letter 'n' following locators refers to notes

Printed and bound in the United States of America